ACHIEVE
The expected standard

Year 6

Mathematics

SATs Question Workbook

Steph King
& Sarah-Anne Fernandes

Rising Stars

Every effort has been made to trace all copyright holders, but if any have been inadvertently overlooked, the Publishers will be pleased to make the necessary arrangements at the first opportunity.

Although every effort has been made to ensure that website addresses are correct at time of going to press, Rising Stars cannot be held responsible for the content of any website mentioned in this book. It is sometimes possible to find a relocated web page by typing in the address of the home page for a website in the URL window of your browser.

Hachette UK's policy is to use papers that are natural, renewable and recyclable products and made from wood grown in sustainable forests. The logging and manufacturing processes are expected to conform to the environmental regulations of the country of origin.

Orders: please contact Bookpoint Ltd, 130 Park Drive, Milton Park, Abingdon, Oxon OX14 4SE. Telephone: (44) 01235 400555.
Email: primary@bookpoint.co.uk

Lines are open from 9 a.m. to 5 p.m., Monday to Saturday, with a 24-hour message answering service. Visit our website at www.risingstars-uk.com for details of the full range of Rising Stars publications.

Online support and queries email: onlinesupport@risingstars-uk.com

ISBN: 978 1 51044 267 2

© Hodder & Stoughton Limited 2018

Reprinted 2019

First published in 2015

This edition published in 2018 by Hodder & Stoughton Limited (for its Rising Stars imprint, part of the Hodder Education Group),
An Hachette UK Company
Carmelite House
50 Victoria Embankment
London EC4Y 0DZ

www.risingstars-uk.com

Impression number 10 9 8 7 6 5

Year 2023 2022 2021 2020 2019

All rights reserved. Apart from any use permitted under UK copyright law, no part of this publication may be reproduced or transmitted in any form or by any means, electronic or mechanical, including photocopying and recording, or held within any information storage and retrieval system, without permission in writing from the publisher or under licence from the Copyright Licensing Agency Limited. Further details of such licences (for reprographic reproduction) may be obtained from the Copyright Licensing Agency Limited, https://www.cla.co.uk/

Authors: Steph King and Sarah-Anne Fernandes

Series Editor: Sarah-Anne Fernandes

Accessibility Reviewer: Vivien Kilburn

Cover design: Burville-Riley Partnership

Illustrations by Ann Paganuzzi

Typeset in India

Printed in Slovenia

A catalogue record for this title is available from the British Library.

Contents

Introduction	4

NUMBER AND PLACE VALUE
Place value of whole numbers	6
Comparing and ordering whole numbers	7
Rounding	8
Place value of decimal numbers	9
Negative numbers	10

NUMBER – ADDITION, SUBTRACTION, MULTIPLICATION AND DIVISION
Addition	11
Subtraction	12
Multiplying and dividing by 10 and 100	13
Multiples and factors	14
Multiplying by larger numbers	15
Square numbers	16
Short division	17
Long division	18
Prime numbers	19

NUMBER – FRACTIONS, DECIMALS AND PERCENTAGES
Fractions of amounts	20
Mixed numbers	21
Equivalent fractions	22
Adding and subtracting fractions	23
Fractions and their decimal equivalents	24
Adding and subtracting decimals	25
Multiplying decimals	26
Percentages as fractions and decimals	27
Finding percentages	28

RATIO AND PROPORTION
Ratio and proportion	29

ALGEBRA
Algebra	30
Sequences	31
Solving equations	32

MEASUREMENT
Length	33
Mass	34
Capacity	35
Money	36
Time	37
Time problems	38
Perimeter	39
Estimating the area of irregular shapes	40
Area by formula	41

GEOMETRY – PROPERTIES OF SHAPES
Drawing lines and angles	42
2-D shapes	43
3-D shapes	44
Angles and degrees	45
Angles in triangles	46

GEOMETRY – POSITION AND DIRECTION
Coordinates	47
Translations	48
Reflections	49
Reflective symmetry	50

STATISTICS
Tables	51
Pictograms	52
Bar charts	53
Pie charts	54
Line graphs	55
Averages	56

Answers	57

INTRODUCTION

Welcome to Achieve Mathematics: The Expected Standard – Question Workbook

In this book you will find lots of practice and information to help you achieve the expected standard in the Key Stage 2 Mathematics tests.

About the Key Stage 2 Mathematics National Tests

The tests will take place in the summer term in Year 6. They will be done in your school and will be marked by examiners – not by your teacher.

There are three papers to the tests:

Paper 1: Arithmetic – 30 minutes (40 marks)
- These questions assess confidence with a range of mathematical operations.
- Most questions are worth 1 mark. However, 2 marks will be available for long multiplication and long division questions.
- It is important to show your working – this may gain you a mark in questions worth 2 marks, even if you get the answer wrong.

Papers 2 and 3: Reasoning – 40 minutes (35 marks) per paper
- These questions test mathematical fluency, solving mathematical problems and mathematical reasoning.
- Most questions are worth 1 or 2 marks. However, there may be one question worth 3 marks.
- There will be a mixture of question types, including multiple-choice, true/false or yes/no questions, matching questions, short responses such as completing a chart or table or drawing a shape, or longer responses where you need to explain your answer.
- In questions that have a method box it is important to show your method – this may gain you a mark, even if you get the answer wrong.

You will be allowed to use a pencil/black pen, an eraser, a ruler, an angle measurer/protractor and a mirror. You **are not allowed** to use a calculator in any of the test papers.

INTRODUCTION

Test techniques

Before the tests

- Try to revise little and often, rather than in long sessions.
- Choose a time of day when you are not tired or hungry.
- Choose somewhere quiet so you can focus.
- Revise with a friend. You can encourage and learn from each other.
- Read the 'Top tips' throughout this book to remind you of important points in answering test questions.
- Keep track of your score using the table on the inside back cover of this book.

During the tests

- READ THE QUESTION AND READ IT AGAIN.
- If you find a question difficult to answer, move on; you can always come back to it later.
- Always answer a multiple-choice question. If you really can't work out an answer, try to think of the most sensible response and read the question again.
- Check to see how many marks a question is worth. Have you written enough to 'earn' those marks in your answer?
- Read the question again after you have answered it. Make sure you have given the correct number of answers within a question, e.g. if there are two boxes for two missing numbers.
- If you have any time left at the end, go back to the questions you have missed.

Where to get help:
- **Pages 6–10** practise number and place value.
- **Pages 11–19** practise number – addition, subtraction, multiplication and division.
- **Pages 20–28** practise number – fractions, decimals and percentages.
- **Page 29** practises ratio and proportion.
- **Pages 30–32** practise algebra.
- **Pages 33–41** practise measurement.
- **Pages 42–46** practise geometry – properties of shapes.
- **Pages 47–50** practise geometry – position and direction.
- **Pages 51–56** practise statistics.
- **Pages 57–64** provide the answers to the questions.

The pencil icon appears next to questions for which you should show your workings.

NUMBER AND PLACE VALUE

Place value of whole numbers

To achieve the expected standard, you need to:
★ know the **place value** of digits in whole numbers up to 1,000,000 and begin to know the place value of digits in whole numbers up to 10,000,000.

1 Write the value of the digit **9** in each of these numbers.

1,904   29,456   4,125,692   1,932,400

Hundreths   *Ten Hundreths*   *Tens*   *Hundreth Thousand* [handwritten answers]

(1 mark)

2 Draw lines to match the numbers with the value of the digit shown.

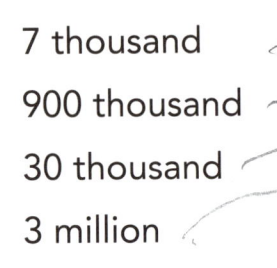

7 thousand   1,233,420
900 thousand   597,302
30 thousand   3,095,799
3 million   979,200

(2 marks)

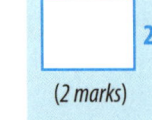

3 Use **all** of the digit cards to make **two** different numbers that can be placed on the shaded part of this number line. You can only use each card once.

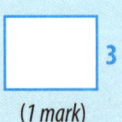

| 0 | 1 | 3 | 4 | 5 | 6 | 7 | 8 |

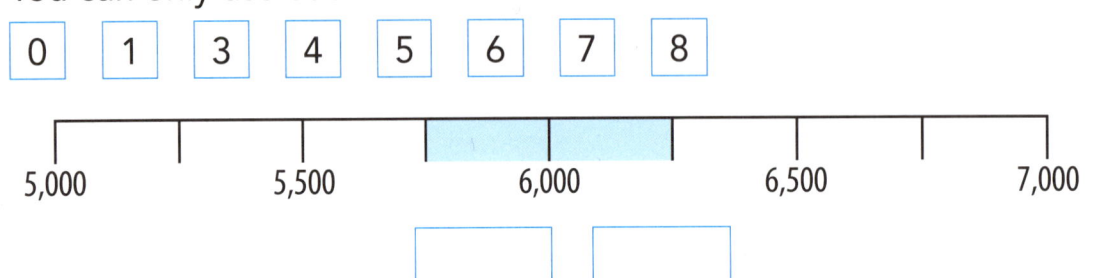

5,000   5,500   6,000   6,500   7,000

(1 mark)

4 Circle the **two** numbers that have a difference of 20,000

134,200   3,463,400   3,263,400   136,200   3,483,400

(1 mark)

5 Asha subtracts two numbers from the value shown on the calculator.

Her calculator display then reads 6,000,803

The first number she subtracted was 400,000

What was the second number?

[Calculator display: 6409803]

(1 mark)

Top tip
- Find the missing values on a number line first.

/ 6

Total for this page

6

Comparing and ordering whole numbers

NUMBER AND PLACE VALUE

To achieve the expected standard, you need to:
★ use place value of digits in whole numbers to compare and order numbers up to 1,000,000
★ begin to work with numbers up to 10,000,000.

1 Write in the missing sign <, > or = to make each statement true.
   a) 34,601 **>** 34,599   b) 709,898 **<** 709,988

(1 mark)

2 Write these masses in order from **heaviest** to **lightest**.

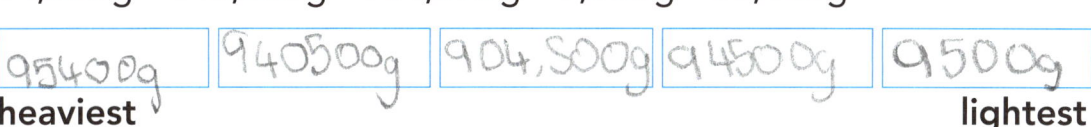

94,500 g   940,500 g   904,500 g   9,500 g   95,400 g

95400g	940500g	904,500g	9450 0g	9500g
heaviest				lightest

(1 mark)

3 Choose a missing digit for each number so that they are in order from **smallest** to **largest**.

Write the digits in the boxes.

2,5**1**0,500   2,519,**2**85   2,**4**05,250   2,60**6**,125

smallest   largest

(1 mark)

4 This table shows the total number of visitors to the zoo each year.

2012	2013	2014	2015	2016	2017
695,099	695,100	607,313	700,995	670,133	617,113

Write the years in the correct order in the table below. One has been done for you.

	Year
Greatest number of visitors	2015
	2013
	2014
	2012
	2016
Fewest number of visitors	2017

(1 mark)

Top tip
- When comparing numbers, write the numbers in place value columns and make sure the columns line up.

/4
Total for this page

7

NUMBER AND PLACE VALUE

Rounding

To achieve the expected standard, you need to:
★ **round** any whole number to the nearest 10; 100; 1,000; 10,000; or 100,000.

1 Write digits in the boxes to make this statement true.

☐78 and 5☐0 both round to 500 when rounded to the nearest hundred.

(1 mark)

2 Round these lengths to the **nearest** 10 cm.

a) 125 cm ☐ cm

b) 99 cm ☐ cm

c) 543 cm ☐ cm

(1 mark)

3 Circle the numbers that round to 1,000 when rounded to the nearest **hundred**.

949   1,050   952   1,045   1,100   1,009

(1 mark)

4 Round each of these numbers as described.

	To the nearest 10	To the nearest 1,000	To the nearest 10,000
45,419			
234,549			
3,126,095			

(2 marks)

5 Ashton thinks of a number and rounds it as shown in the table below.

	To the nearest 1,000	To the nearest 100	To the nearest 10
Ashton's number	6,000	5,700	5,750

Find the smallest number and the largest number that Ashton could have used.

a) Smallest: ☐

b) Largest: ☐

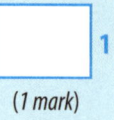

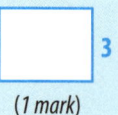

(2 marks)

/7

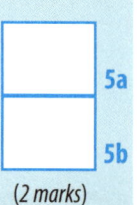

Total for this page

NUMBER AND PLACE VALUE

Place value of decimal numbers

To achieve the expected standard, you need to:
★ know the place value of digits in numbers with up to two decimal places.

1 3 ÷ ☐ = 0.3

(1 mark)

2 Fill in the missing terms in this sequence.

2.03   1.98   ☐   1.88   ☐   ☐

(1 mark)

3 Pete counts on five steps of **0.09** from 2.5

Write each of the numbers he lands on.

2.5   ☐   ☐   ☐   ☐   ☐

(1 mark)

4 Jennie writes down a number.

It has 4 hundredths, no tenths and 8 thousandths.

What is Jennie's number? ☐

(1 mark)

5 Here is a set of decimal numbers:

0.6   6.66   6   0.06

Ben says 6 is the smallest number.

Explain why Ben is incorrect.

_____

_____

(1 mark)

6 Write these numbers in order of size, starting with the smallest.

| 7.02 | 27.0 | 0.72 | 2.70 | 7.2 |

☐   ☐   ☐   ☐   ☐

smallest   largest

(1 mark)

Top tip
- When comparing numbers, write the numbers in place value columns and make sure the columns line up.

/ 6

Total for this page

NUMBER AND PLACE VALUE

Negative numbers

To achieve the expected standard, you need to:
★ use negative numbers in practical contexts
★ calculate intervals across zero.

1 Fill in the missing numbers in this sequence.

7   4   **1**   –2   **–5**

↳ Difference is 3.

(1 mark)

2 Sam parks his car on **Level –2** of the town car park.

How many levels must he go up to reach the exit to the shops on **Level 3**? **5**

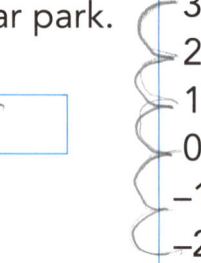

(1 mark)

3 The thermometer shows the temperature recorded in France on a day in June.

The temperature recorded on a day in December was 32°C **lower**.

What was the temperature in December? **7** °C

²⁸³²
 − 25
 ─────
   7

(1 mark)

4 Paulo picks two number cards. One of the numbers is negative and one of the numbers is positive.

The difference between the two numbers is 11

a) Find **two** different pairs of numbers he could pick.

☐ and ☐

☐ and ☐

(1 mark)

b) Jan said that one of the numbers could be 11

Is Jan correct? Circle your answer. YES / NO

Explain your answer.

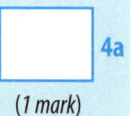

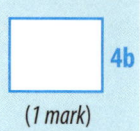

_____

_____

_____

/5

Total for this page

10

NUMBER – ADDITION, SUBTRACTION, MULTIPLICATION AND DIVISION

Addition

To achieve the expected standard, you need to:
★ secure adding whole numbers up to four digits
★ add whole numbers with more than four digits
★ use formal written methods.

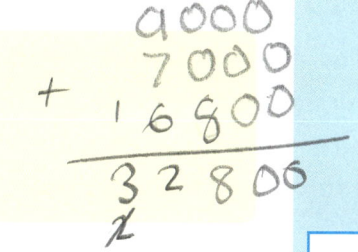

1 Write the missing digits. 9,000 + 7,000 + |16 000| = 16,800

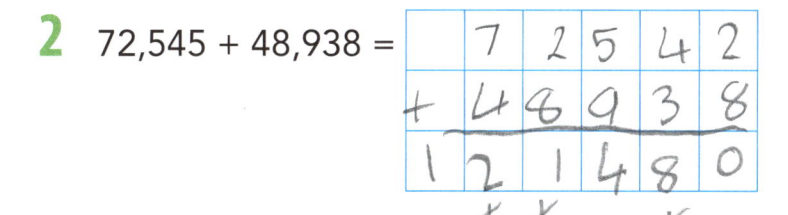

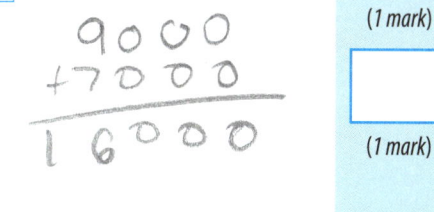

(1 mark)

2 72,545 + 48,938 =

7	2	5	4	2
+4	6	9	3	8
1 2	1	4	8	0

(1 mark)

3 a) How many points do the Red and Green teams score altogether?

Red 7,409
Blue 6,598
Green 8,067
Points scored

|15476|

(1 mark)

b) What is the total number of points scored by all three teams?

|22074|

(1 mark)

4 Write the missing digits to make the addition correct.

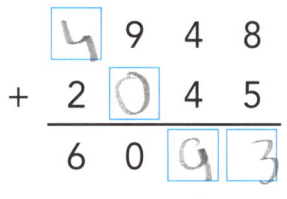

```
  4 9 4 8
+ 2 0 4 5
  6 0 9 3
```

(1 mark)

5 A marathon run raised £235,748 for a charity in 2016
In 2017 the run raised £79,200 more than in 2016
What is the total amount of money raised so far?   £ 314948

2	3	5	7	4	8
+	7	9	2	0	0
3	1	4	9	4	8

(2 marks)

Top tip
• Write down any numbers that you need to carry in the correct column so you remember to include them.

/7

Total for this page

11

NUMBER – ADDITION, SUBTRACTION, MULTIPLICATION AND DIVISION

Subtraction

To achieve the expected standard, you need to:
- secure subtracting whole numbers up to four digits
- subtract whole numbers with more than four digits
- use formal written methods.

1 34,562 − 23,638 = 10924

```
  3³ ⁴4 ⁵5 ⁶6¹ 2
−   2  3  6  3  8
─────────────────
     1  0  9  2  4
```

(1 mark)

2 Find the difference between 450,000 and 290,000   160000

```
  ³4⁵50 000
−  2 90 000
───────────
   1 60 000
```

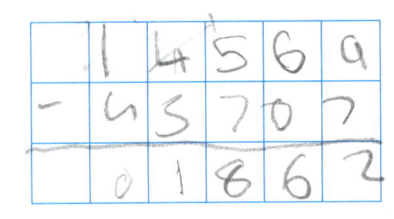

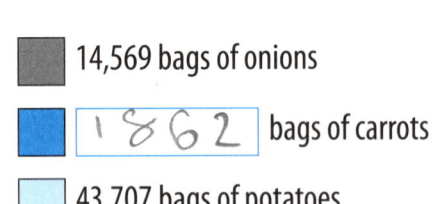

(1 mark)

3 How many bags of carrots were sold?

```
  1  4  5  6  9
−    4  5  7  0  7
───────────────────
  0  1  8  6  2
```

Vegetables sold

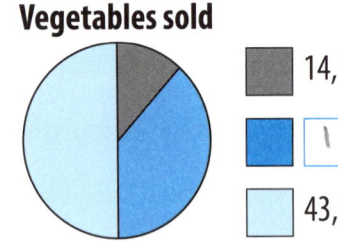

- 14,569 bags of onions
- 1862 bags of carrots
- 43,707 bags of potatoes

(1 mark)

4 The hospital is collecting money for charity. Their target is to collect £15,550.

The doctors have donated £2,759. The patients have donated £5,764. The coffee shop in the hospital has raised £1,905.

How much **more** money do they need to reach their target?

£5122

```
   15550
−   2759
   5764
─────────
      05
```

(2 marks)

5 Write the missing digits to make the subtraction correct.

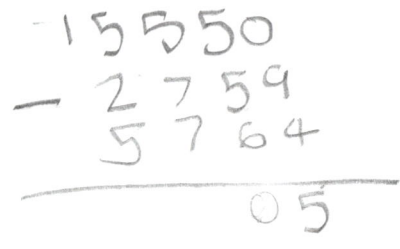

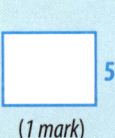

```
   4 0 ☐ 8 6
 − 2 ☐ 6 3 ☐
   ─────────
   1 6 2 ☐ 9
```

(1 mark)

Top tip
- Remember to check your answer using the inverse operation: addition.

/ 6

Total for this page

NUMBER – ADDITION, SUBTRACTION, MULTIPLICATION AND DIVISION

Multiplying and dividing by 10 and 100

To achieve the expected standard, you need to:
★ multiply and divide whole numbers and decimal numbers with up to two decimal places by 10 or 100.

1 Fill in the missing numbers.

a) 235 × 100 = `23500`   b) 745 ÷ `100` = 7.45

c) `0.99` × 100 = 99

(1 mark)

2 Draw lines to match each calculation with the correct answer.

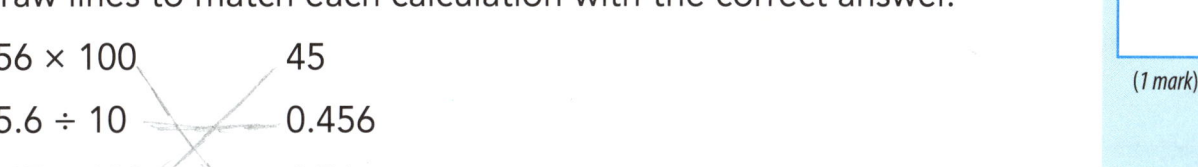

456 × 100   45
45.6 ÷ 10   0.456
0.45 × 100   4.56
4.56 ÷ 10   45,600

(1 mark)

3 Asha draws a plan of her new sandpit.

A length of 100 cm on the real sandpit is drawn as 10 cm on her plan.

Fill in the missing dimensions on Asha's plan.

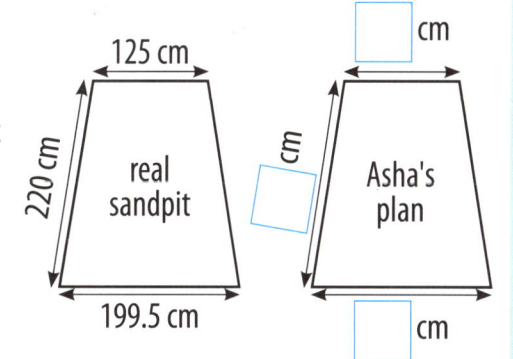

real sandpit: 125 cm (top), 220 cm (side), 199.5 cm (bottom)

Asha's plan: ☐ cm (top), ☐ cm (side), ☐ cm (bottom)

(2 marks)

4 Use the clues to find the mass of the other parcels.

- Parcel C is **100 times lighter** than parcel A.
- Parcel A is **10 times heavier** than parcel B.

Parcel A is ☐ kg and parcel B is ☐ kg.

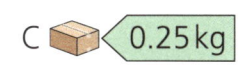

A   B   C ◀ 0.25 kg

(1 mark)

Top tip
- Read the question carefully and decide which operation you need to use.

/ 5

Total for this page

NUMBER – ADDITION, SUBTRACTION, MULTIPLICATION AND DIVISION

Multiples and factors

To achieve the expected standard, you need to:
★ recognise and use **multiples** and **factors**.

1 Circle **all** the multiples of **11** listed below.

77   220   144   177   1,100   111

(1 mark)

2 Sophie is thinking of an **odd** number.

It is a factor of 24 **and** a factor of 15

What is the number? ☐

(1 mark)

3 Draw lines to match the multiples with their factors.

One of the factors is missing. It is **not** the factor 1

Write a possible missing factor in the box.

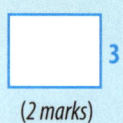

Factor   Multiple

☐
              42
5
              24
12
              40
8

(2 marks)

4 Complete the wheel for the number 32 by writing all the factors of 32 in the spaces on the wheel.

(32)

(1 mark)

5 Write three factors of 30 that are not factors of 15.

☐ ☐ ☐

(1 mark)

/ 6

Total for this page

NUMBER – ADDITION, SUBTRACTION, MULTIPLICATION AND DIVISION

Multiplying by larger numbers

To achieve the expected standard, you need to:
★ draw upon multiplication facts up to 12 × 12 and place value to:
  – multiply numbers with up to four digits by a one-digit number using short multiplication
  – multiply numbers with up to four digits by a two-digit number using the formal long multiplication method and become more confident with larger numbers.

1 3,609 × 7 =

(1 mark)

2
```
    3 8 5
  ×   1 3
```

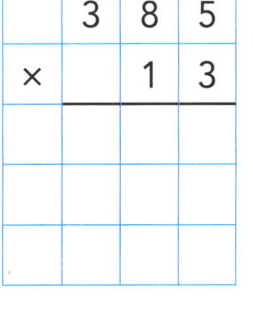

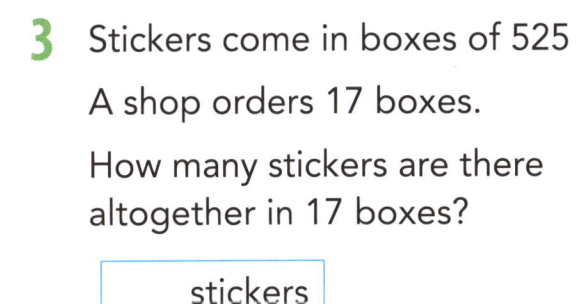

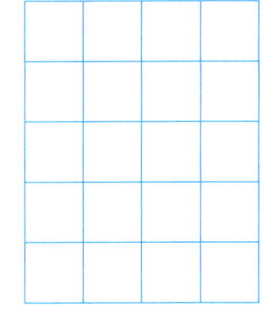

(2 marks)

3 Stickers come in boxes of 525

A shop orders 17 boxes.

How many stickers are there altogether in 17 boxes?

☐ stickers

(1 mark)

4 | 3,247 | | | | |

This bar is divided into equal parts.

Jake writes ☐ × ☐ = ☐

to help him find the total value of the bar.

Complete Jake's calculation.

(2 marks)

5 A shop orders 52 boxes of oranges.

Each box contains 24 bags of oranges. Each bag contains 12 oranges.

How many oranges does the shop order in total? ☐ oranges

(2 marks)

Top tip
- Always show your method because you could gain a mark even if your answer is incorrect.

/ 8

Total for this page

NUMBER – ADDITION, SUBTRACTION, MULTIPLICATION AND DIVISION

Square numbers

To achieve the expected standard, you need to:
★ recognise and use **square numbers** up to 144.

1 Use two different **square numbers** to make this statement true.

☐ − 9 = ☐

(1 mark)

2 Jack only saves **1 pence** coins. He makes different **squares** using the coins he has saved in his money box.

This is Jack's **4p** square.

What is the value of the **largest** square that Jack can make? ☐

(1 mark)

3 Find **two** different ways to make this calculation true.

square number   square number   prime number

☐ + ☐ = ☐

☐ + ☐ = ☐

(1 mark)

4 Paul says that this square has sides 9 cm long, because 9 × 4 is 36

Amy says the sides are 6 cm long.

Who is correct? ☐

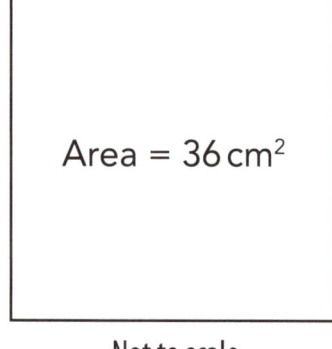

Area = 36 cm²

Not to scale

(1 mark)

5 Ben says there are two square numbers between 80 and 110.

Do you agree? YES / NO

Explain your thinking.

_____

_____

/5

Total for this page

NUMBER – ADDITION, SUBTRACTION, MULTIPLICATION AND DIVISION

Short division

To achieve the expected standard, you need to:
★ use the formal written method of short division to divide numbers with up to four digits by a one-digit number
★ divide whole numbers mentally, drawing upon multiplication facts and place value, and begin to use these facts to work with larger numbers.

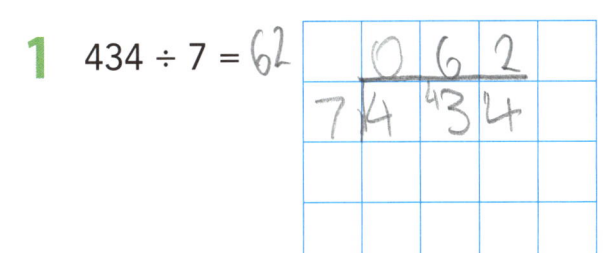

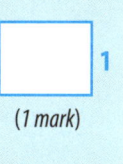

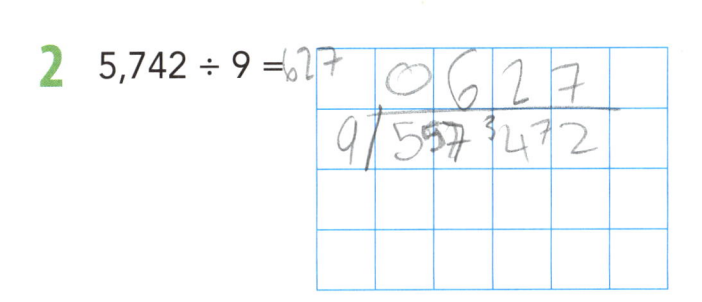

1 434 ÷ 7 = 62

(1 mark)

2 5,742 ÷ 9 = 627

(1 mark)

3

| 5,304 |

| ? | | | | | |

The value of the whole bar is 5,304

Write the value of the shaded part. ☐

(1 mark)

4 Each table has 8 seats.

How many tables are needed to seat 349 people?

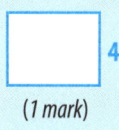

☐ tables

(1 mark)

5 The total capacity of six different containers is **2,865 millilitres**.

What is the average (mean) capacity of a container? ☐ ml

(1 mark)

Top tip
- In a problem like question 4, remember that one table may not be completely full.
- You must show a remainder as a fraction or a decimal when calculating the mean as an average.

/ 5

Total for this page

NUMBER – ADDITION, SUBTRACTION, MULTIPLICATION AND DIVISION

Long division

To achieve the expected standard, you need to:
★ become more confident using long division to divide numbers with up to four digits by a two-digit whole number.

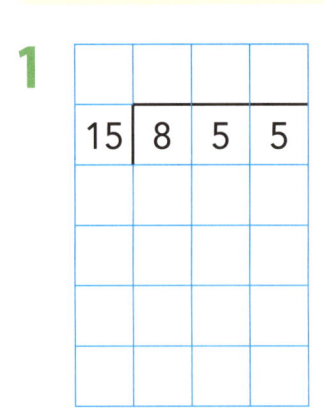

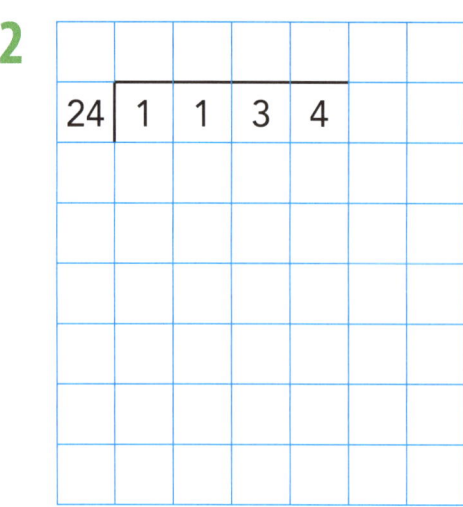

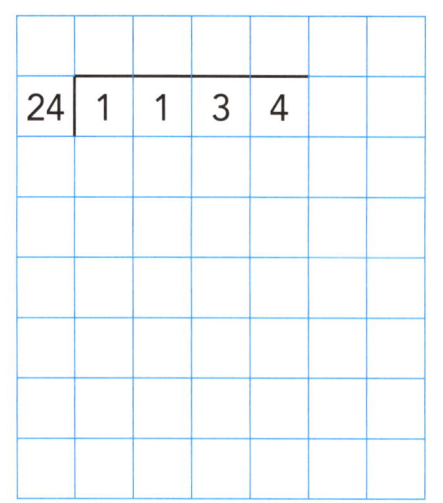

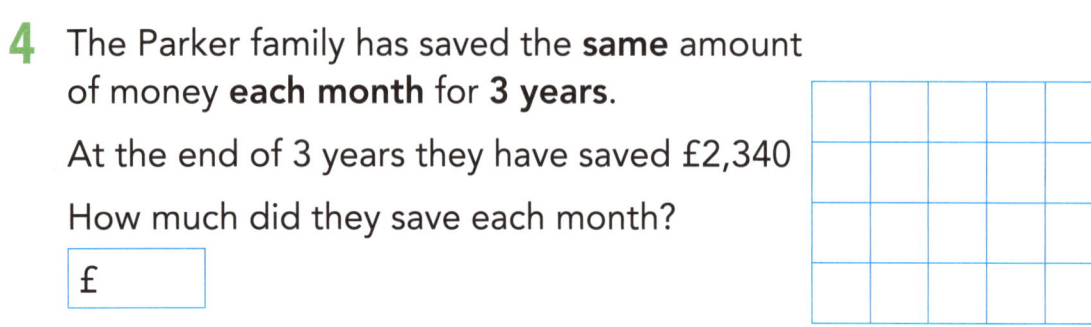

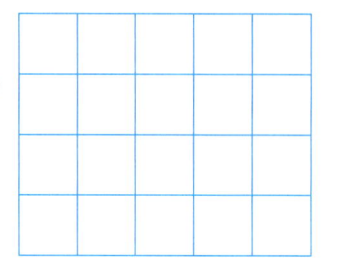

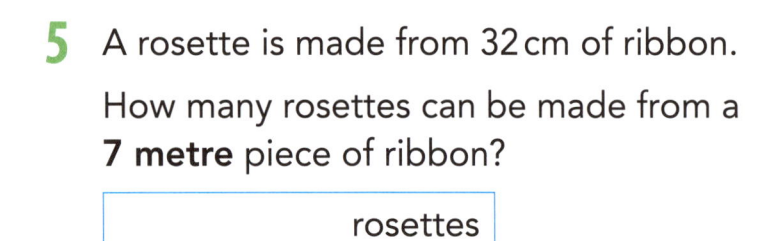

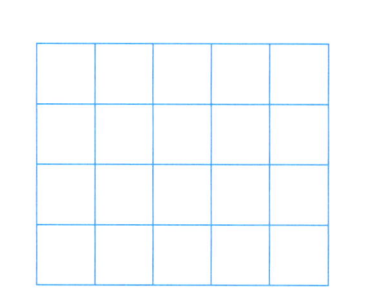

1 15) 8 5 5

2 24) 1 1 3 4

(2 marks)

(2 marks)

3 A car uses **18 litres** of petrol to travel a distance of **594 kilometres**. How many kilometres does it travel **per litre** of petrol? ☐ km

(1 mark)

4 The Parker family has saved the **same** amount of money **each month** for **3 years**.

At the end of 3 years they have saved £2,340

How much did they save each month?

£ ☐

(2 marks)

5 A rosette is made from 32 cm of ribbon.

How many rosettes can be made from a **7 metre** piece of ribbon?

☐ rosettes

(2 marks)

Top tip
- Always show your method because you could gain a mark even if your answer is incorrect.

/9

Total for this page

NUMBER – ADDITION, SUBTRACTION, MULTIPLICATION AND DIVISION

Prime numbers

To achieve the expected standard, you need to:
★ recognise and use **prime numbers** less than 20.

1 Complete the list of prime numbers.

☐ 3 5 ☐ ☐ 13 ☐ 19

(1 mark)

2 Which prime number is a factor of both 9 and 18? ☐

(1 mark)

3 Izi picks one **odd** prime number and one **even** prime number.

The difference between the two prime numbers is 15

Which two prime numbers does she pick?

☐ and ☐

(1 mark)

4 Callum is thinking of a number.

It is a factor of 30 and it is prime. What could it be? ☐

(1 mark)

5 42 children get into groups of an equal size.

The number of children in each group is **prime**.

The number of children in each group is ☐ or ☐ or ☐.

(1 mark)

6 The **sum** of three **prime** numbers is **26**

The three numbers are:

☐ + ☐ + ☐

or ☐ + ☐ + ☐

or ☐ + ☐ + ☐

(2 marks)

Top tip

- Check you have chosen numbers that match all the criteria in a question.
- Remember that **1** is **not** a prime number. There is only one **even** prime number (2).

/7

Total for this page

19

NUMBER – FRACTIONS, DECIMALS AND PERCENTAGES

Fractions of amounts

To achieve the expected standard, you need to:
★ calculate simple fractions of whole numbers and quantities.

1 a) $\frac{1}{8}$ of 40 = ☐   b) $\frac{3}{8}$ of 40 = ☐

(1 mark)

2 Megan has $\frac{3}{4}$ of £120 and Ali has $\frac{3}{8}$ of £160

How much **more** money does Megan have? ☐

(1 mark)

3 The **whole** container holds **3.6 litres** of water.

How much water is shown here? ☐ litres

(1 mark)

4 Three schools raised £4,800 in total.
a) ☐ was raised by Oak School.
b) ☐ was raised by Ash School.

(1 mark) 4a

(1 mark) 4b

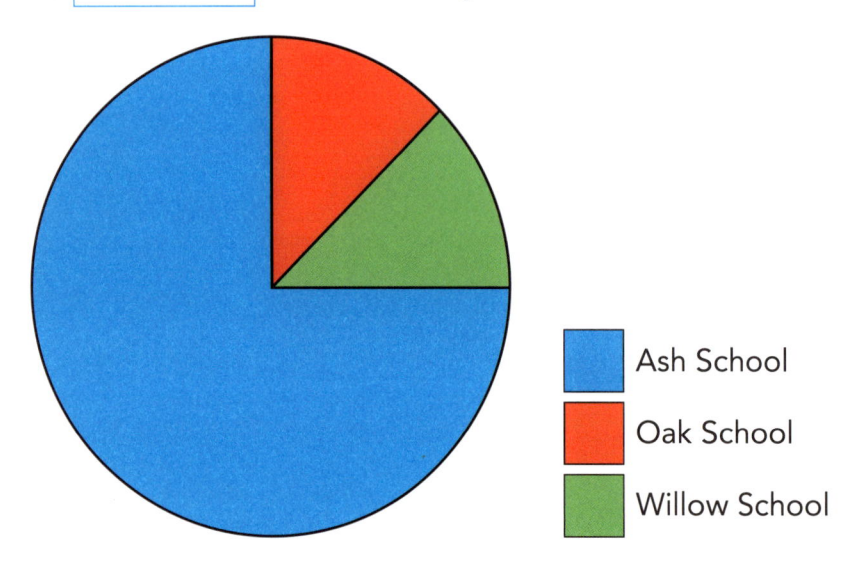

- Ash School
- Oak School
- Willow School

5 £45 is $\frac{☐}{5}$ of £75

(1 mark)

6 On Monday, Ben read $\frac{1}{3}$ of his book.

On Tuesday, he read the **other** 72 pages to finish the book.

How many pages are there in Ben's book? ☐

(2 marks)

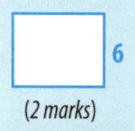

Top tip
- To find fractions of amounts always divide by the denominator and then multiply by the numerator.

/8

Total for this page

NUMBER – FRACTIONS, DECIMALS AND PERCENTAGES

Mixed numbers

To achieve the expected standard, you need to:
★ recognise **mixed numbers** and **improper fractions** and convert from one to the other.

1 What is $6\frac{2}{3}$ as an **improper fraction**?

(1 mark)

2 Rewrite this sequence of improper fractions as **mixed numbers**.

$\frac{5}{4}$ $\frac{7}{4}$ $\frac{9}{4}$ $\frac{11}{4}$ $\frac{13}{4}$

(1 mark)

3 Write the answer to this calculation as a **mixed number**.

$\frac{7}{8} + \frac{5}{8} =$

(1 mark)

4 Complete the empty boxes on this number line.

mixed numbers $3\frac{2}{5}$

improper fractions $\frac{22}{5}$

(2 marks)

5 Circle **all** the improper fractions that are **equivalent** to 6

$\frac{12}{6}$ $\frac{30}{5}$ $\frac{18}{3}$ $\frac{60}{12}$ $\frac{54}{9}$

(1 mark)

6 Write the amount shaded as a mixed number and as an improper fraction.

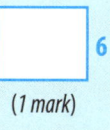

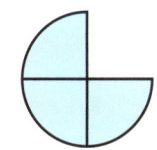

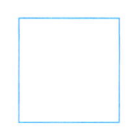

(1 mark)

Top tip
- Remember that $\frac{3}{3}$, $\frac{4}{4}$ and $\frac{6}{6}$ are all equivalent to 1.

/7

Total for this page

NUMBER – FRACTIONS, DECIMALS AND PERCENTAGES

Equivalent fractions

To achieve the expected standard, you need to:
★ recognise and use **equivalent fractions**
★ find equivalent fractions with lower denominators
★ rewrite a pair of fractions so they share the same denominator.

1 Draw lines to join the equivalent fractions.

$\frac{2}{3}$ $\frac{75}{100}$

$\frac{3}{4}$ $\frac{3}{18}$

$\frac{2}{5}$ $\frac{8}{12}$

$\frac{1}{6}$ $\frac{40}{100}$

(1 mark)

2 $\frac{3}{5} = \frac{15}{\Box}$

(1 mark)

3 Simplify the fractions $\frac{5}{20}$ and $\frac{9}{12}$ so they have the **same** denominator.

(1 mark)

4 Circle the fractions of a metre that are **not** equivalent to 40 cm.

$\frac{4}{10}$ m $\frac{4}{100}$ m $\frac{2}{5}$ m $\frac{40}{100}$ m $\frac{1}{4}$ m

(1 mark)

5 Rewrite each of these fractions so that they all have the **same** denominator.

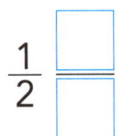

$\frac{1}{2}\frac{\Box}{\Box}$ $\frac{2}{5}\frac{\Box}{\Box}$ $\frac{3}{4}\frac{\Box}{\Box}$

(1 mark)

6 Shade the diagram to show a fraction that is equivalent to $\frac{15}{24}$

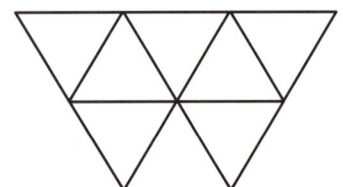

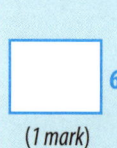

(1 mark)

Top tip
• Use your multiplication facts to help find equivalent fractions.

/6
Total for this page

NUMBER – FRACTIONS, DECIMALS AND PERCENTAGES

Adding and subtracting fractions

To achieve the expected standard, you need to:
★ add and subtract fractions with the same denominator, using mixed numbers where appropriate for the context
★ add and subtract fractions with denominators that are **multiples** of the same number, and become more confident with more complex calculations.

1 $1\frac{4}{5} - \frac{2}{5} = \boxed{}$

(1 mark)

2 $1\frac{3}{5} - \frac{4}{5} = \boxed{}$

(1 mark)

3 $\boxed{} = \frac{2}{3} + \frac{5}{6}$

(1 mark)

4 $\frac{3}{4} + \boxed{} + \frac{1}{2} = 2$

(1 mark)

5 Draw lines to match these calculations with their answers.

$1\frac{5}{8} - \frac{3}{4}$ $3\frac{1}{2}$

$1\frac{3}{4} - 1\frac{1}{12}$ 3

$1\frac{3}{5} + \frac{7}{10} + \frac{7}{10}$ $\frac{7}{8}$

$2\frac{5}{6} + \frac{2}{3}$ $\frac{2}{3}$

(2 marks)

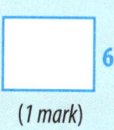

6 There is $\frac{3}{4}$ litre of water in a jug.

Jack pours in a further $\frac{3}{8}$ litre of water.

How much water is in the jug altogether?

Write your answer as a mixed number. $\boxed{}$ litres

(1 mark)

/7

Total for this page

NUMBER – FRACTIONS, DECIMALS AND PERCENTAGES

Fractions and their decimal equivalents

To achieve the expected standard, you need to:
★ recognise and write decimal equivalents for $\frac{1}{4}$, $\frac{1}{2}$, $\frac{3}{4}$ and any number of fifths, tenths or hundredths and read and write decimals as fractions
★ recognise and use thousandths and relate them to tenths, hundredths and decimal equivalents.

1 Fill in the empty boxes on this number line.

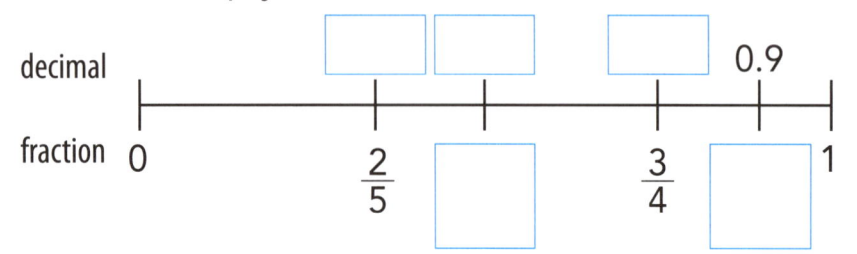

decimal [ ] [ ] [ ] 0.9

fraction 0 $\frac{2}{5}$ [ ] $\frac{3}{4}$ [ ]   1

(2 marks)

2 $0.36 = \dfrac{36}{\boxed{}}$

(1 mark)

3 Circle **all** the fractions that are equivalent to **0.25**

$\dfrac{25}{100}$ $\dfrac{2}{5}$ $\dfrac{5}{20}$ $\dfrac{1}{4}$ $\dfrac{4}{12}$

(1 mark)

4 Write each of the amounts of ingredients for this recipe as a **decimal**.

One has been done for you.

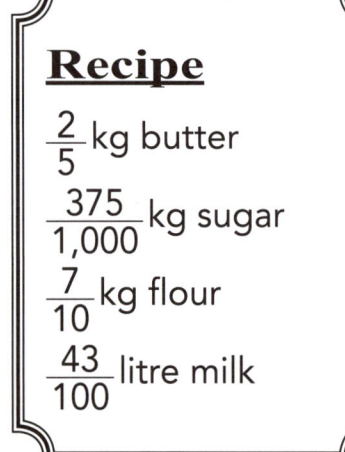

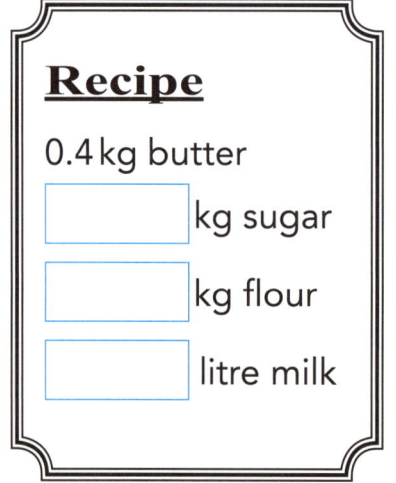

Recipe
$\frac{2}{5}$ kg butter
$\frac{375}{1,000}$ kg sugar
$\frac{7}{10}$ kg flour
$\frac{43}{100}$ litre milk

Recipe
0.4 kg butter
[ ] kg sugar
[ ] kg flour
[ ] litre milk

(1 mark)

Top tip
- Remember, fractions can be converted into decimals by using division (e.g. $\frac{7}{10}$ can be converted using 7 ÷ 10).

/ 5

Total for this page

NUMBER – FRACTIONS, DECIMALS AND PERCENTAGES

Adding and subtracting decimals

To achieve the expected standard, you need to:
★ add and subtract decimal numbers that have the same number of decimal places.

1 a) 64.29 – 20.14 = ☐

b) 7 = 10.9 – ☐

2 Lexi buys these two games.

a) How much does she spend altogether? ☐

b) Work out how much change she will get from £40 ☐

£15.49

£19.75

3 Find **two** different pairs of numbers that total **6**

1.7   2.46   3.64   4.3   3.54   5.3

☐ and ☐ total 6

☐ and ☐ total 6

4 Ollie jumps 1.26 m. Jahla jumps 0.85 m further.

How far does Jahla jump? ☐ m

5

Container A	Container B	Container C	Container D
2.75 litres	4 litres	2.79 litres	8.32 litres

The **total** water in **two** of these containers is 1.53 litres **less than** the amount in **Container D**.

The **two** containers are ☐ and ☐.

Top tip
- Line up the decimal points when adding and subtracting.

(1 mark) 1
(1 mark) 2a
(1 mark) 2b
(1 mark) 3
(1 mark) 4
(1 mark) 5

/ 6

Total for this page

NUMBER – FRACTIONS, DECIMALS AND PERCENTAGES

Multiplying decimals

To achieve the expected standard, you need to:
★ multiply a one-digit decimal number by a one-digit number.

1 ☐ × 9 = 6.3

(1 mark)

2 ☐ = 0.8 × 8

(1 mark)

3 0.05 × ☐ = 0.1

(1 mark)

4 Kai multiplied **0.7** by a one-digit number.

Circle the number that **cannot** be the answer to his calculation.

1.4   3.5   4.7   2.1   5.6

(1 mark)

5 Complete the boxes to make these multiplications correct.

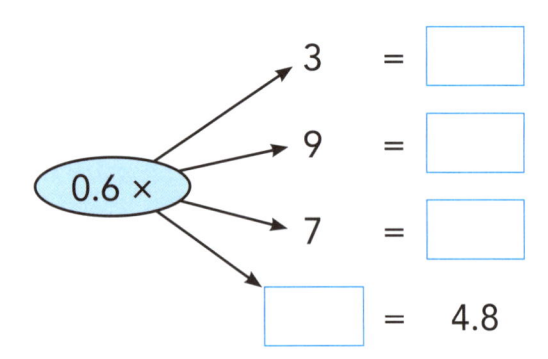

0.6 × 3 = ☐
0.6 × 9 = ☐
0.6 × 7 = ☐
0.6 × ☐ = 4.8

(2 marks)

6 ☐ kg × 5 = 24 kg ÷ 6

(1 mark)

7 Calculate the area of this rectangle. ☐ m²

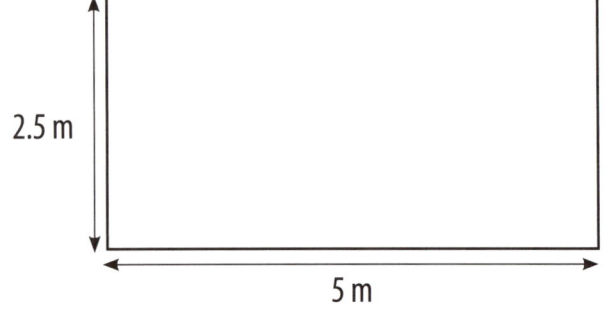

2.5 m
5 m

(1 mark)

Top tip
- Remember to use multiplication facts and place value to help.

/ 8

Total for this page

NUMBER – FRACTIONS, DECIMALS AND PERCENTAGES

Percentages as fractions and decimals

> To achieve the expected standard, you need to:
> ★ recognise and use the equivalences between simple fractions, decimals and **percentages**
> ★ become more confident with calculating other decimal fraction equivalents.

1 Complete this table of equivalent fractions, decimals and percentages.

Fraction	$\frac{3}{10}$			$\frac{4}{5}$
Decimal		0.65		
Percentage			5%	

(2 marks)

2 Circle the amount that is **not** equivalent to $\frac{1}{4}$.

1.4   25% $\frac{3}{12}$ 0.25   14%   2.5

(1 mark)

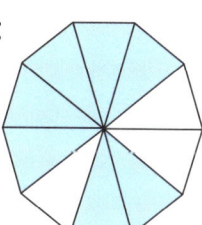

3 The shaded part of this shape can be described as:

☐/☐ or ☐% or ☐.

(1 mark)

4 Fill in the empty boxes on this number line.

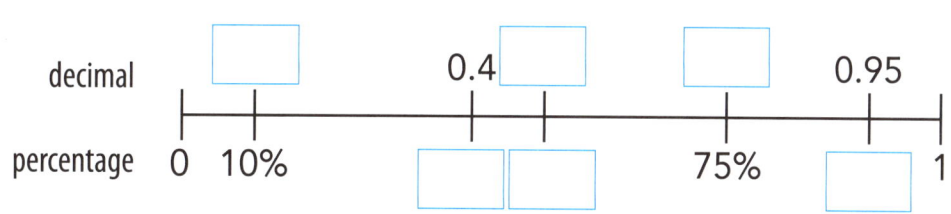

decimal: ☐   0.4   ☐   0.95
percentage: 0   10%   ☐ ☐   75%   ☐   1

(2 marks)

5 George and Alice are sharing a pizza.

George wants to eat **25%** of the pizza.

Alice wants to eat $\frac{2}{5}$ of the pizza.

George thinks that he will be eating the **same** amount of pizza as Alice.

a) Explain why George is incorrect. _____

_____

b) Write Alice's fraction of the pizza as a **decimal**. ☐

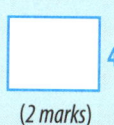

(1 mark) 5a

(1 mark) 5b

/ 8

Total for this page

NUMBER – FRACTIONS, DECIMALS AND PERCENTAGES

Finding percentages

To achieve the expected standard, you need to:
★ find simple percentages of whole numbers and quantities.

1 a) 50% of ☐ = 100   b) 25% of ☐ = 100

(1 mark)

2 20% of 120 = ☐

(1 mark)

3 25% of ☐ = £10

(1 mark)

4 Use the symbol <, > or = to make this statement true.

50% of 300 ☐ 75% of 200

(1 mark)

5 This chart shows the number of visitors to a park.

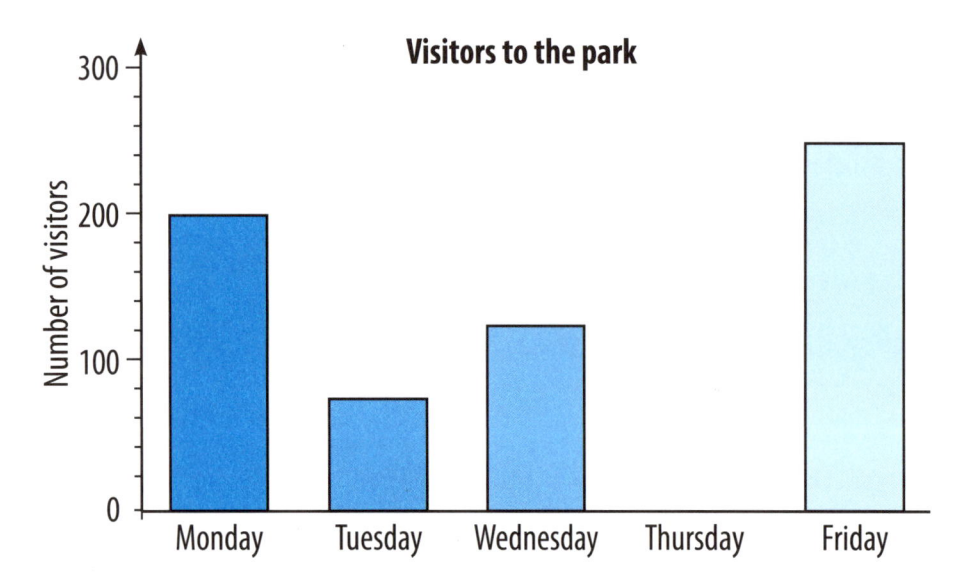

Visitors to the park

(Bar chart: Monday = 200, Tuesday ≈ 75, Wednesday ≈ 125, Thursday = (blank), Friday = 250. Y-axis: Number of visitors, 0 to 300.)

a) The number of visitors on Thursday is 40% of the number of visitors on Monday.

Draw the bar for Thursday on the chart.

(2 marks)

b) 30% of the visitors on Friday were children.

How many adults visited the park on Friday? ☐ adults

(1 mark)

Top tip

- Finding 20% does not mean divide by 20.
- Check the scale of any bar chart first.

/7

Total for this page

Ratio and proportion

To achieve the expected standard, you need to:
★ use simple **ratio** to compare quantities
★ estimate the distance on a map using a simple scale.

1 In Class 2W there are 3 girls to every 2 boys.

There are 18 girls in the class.

How many boys are there? ☐ boys

(1 mark)

2 A recipe uses 100 ml milk for every 250 g flour.

How much milk is needed for **1 kg** of flour? ☐ ml

(1 mark)

3 What is the real distance between Ashley Town and Mason Park?

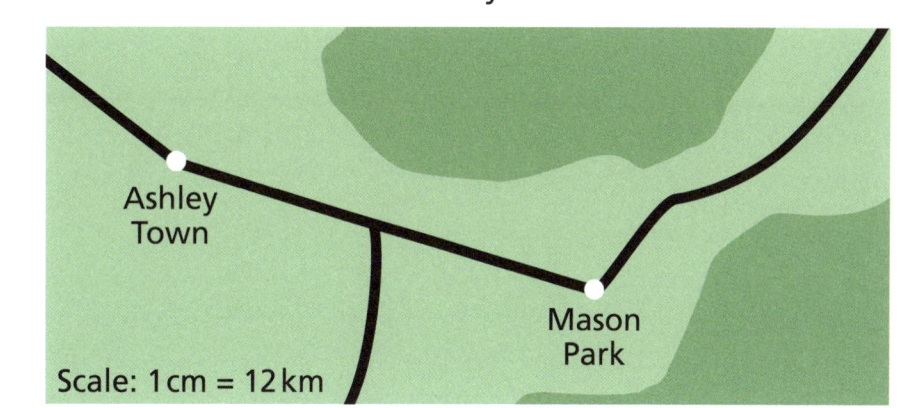

☐ km

Ashley Town

Mason Park

Scale: 1 cm = 12 km

(2 marks)

4 In a group of 20 people, 3 people are left-handed.

What **proportion** of the group is **not** left-handed? ☐

(1 mark)

5 A total of 160 points was scored in a quiz.

The Red team scored 5 points **for every** 3 points scored by the Blue team.

How many points were scored by each team?

Red team ☐

Blue team ☐

(2 marks)

Top tip

- Use your knowledge of multiplication facts to solve direct proportion problems.
- Use factors to help simplify ratios.

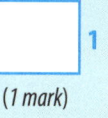

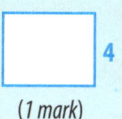

/7

Total for this page

Algebra

To achieve the expected standard, you need to:
★ use simple formulae expressed in words.

1 Use the rule to find the missing numbers.

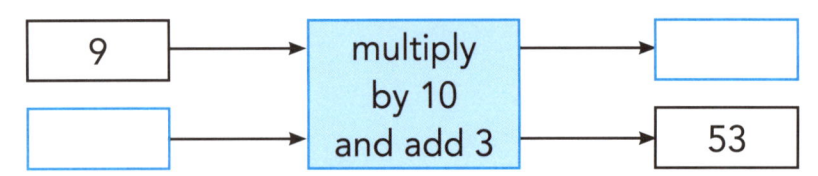

9 → multiply by 10 and add 3 → ☐

☐ → multiply by 10 and add 3 → 53

2 Use the rule below to fill in the empty boxes.

Double a number and subtract 12

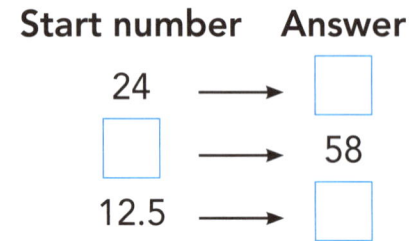

Start number	Answer
24 →	☐
☐ →	58
12.5 →	☐

3 What formula can you write to help James find the area of **any** rectangle?

_____

_____

4 a) Describe the formula needed to calculate missing angle *a* in this triangle.

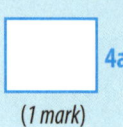

(Triangle with angles 100°, 35°, and *a*)

_____

_____

b) Angle *a* = ☐ °

5 a) *n* = 8

What is 4*n* +16? ☐

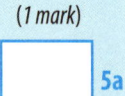

b) 4*q* + 10 = 90

Work out the value of *q*. ☐

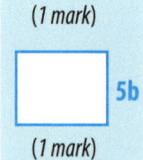

> ⭐ **Top tip**
> • You may want to use brackets to show which part of the calculation to do first.

Sequences

To achieve the expected standard, you need to:
★ count forwards or backwards in steps of any whole number with one significant figure
★ generate, describe and complete linear number **sequences**.

1 A sequence **starts** at **–15**; the rule is **add 7**

Write the next five terms in the sequence.

–15 ☐ ☐ ☐ ☐ ☐

(1 mark)

2 Write the rule for this sequence.

Fill in the missing terms.

The rule is _____

Start
212
End   192
92   What is the rule?   ☐
☐   152

(1 mark)

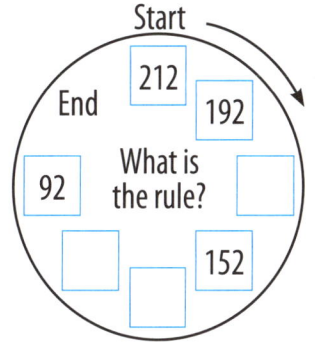

3 Find the rule and the missing terms in this sequence.

248   ☐   128   ☐   8

The rule is _____

(1 mark)

4 The jumps on this number line are of an equal size.

What is the missing value?

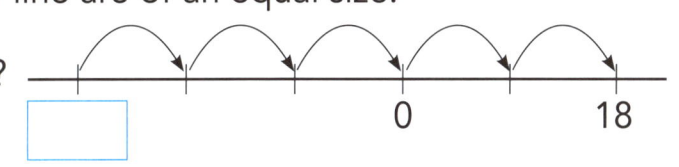

☐ ———————— 0 ———— 18

(1 mark)

5 A sequence starts at 125

The rule is add 700

Will the number 7,025 be in the sequence?

Circle your answer. YES / NO

Explain your answer.

_____

_____

(1 mark)

Top tip

• Look at the number patterns. Think about what stays the same and what changes.

/ 5

Total for this page

ALGEBRA

Solving equations

To achieve the expected standard, you need to:
★ find possible values in missing-number problems and equations involving one or two unknowns.

1 Write numbers in the boxes to make this statement true.

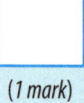

☐ + 15 = ☐ × 3

(1 mark)

2 Two radio programmes last a total of 60 minutes.

One programme is 10 minutes longer than the other.

What is the length of each programme?

☐ and ☐

(1 mark)

3 Aisha thinks of a pair of **factors** of 24

The sum of the two factors is **odd**.

What are the numbers?

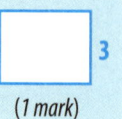

☐ and ☐

  or

☐ and ☐

(1 mark)

4 The pictures show a set of balancing scales and some parcels.

What is the **mass** of each parcel?

Parcel A = ☐ g

Parcel B = ☐ g

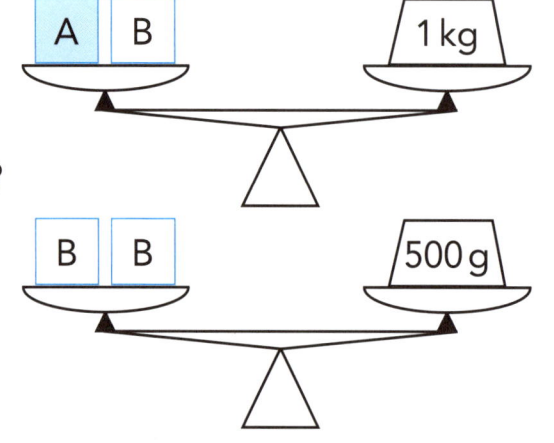

(1 mark)

Top tip
- Use inverse operations to help solve missing number problems.

/ 4

Total for this page

MEASUREMENT

Length

To achieve the expected standard, you need to:
* use, read, write and convert between metric units of length
* use all four operations to solve problems involving length
* use decimals to two decimal places.

1 2 m + ☐ cm = 525 cm

☐ 1
(1 mark)

2 How many **25 centimetre** lengths can be cut from a **4.5 metre** rope? ☐

☐ 2
(1 mark)

3 Draw lines to match the equivalent lengths.

38 cm	38 mm
3.8 cm	380 cm
3.8 m	0.38 m

☐ 3
(1 mark)

4 A car travels a distance of 5.6 kilometres.
How many **metres** is 5.6 kilometres? ☐ m

☐ 4
(1 mark)

5 What is the **perimeter** of the pitch in kilometres? ☐ km

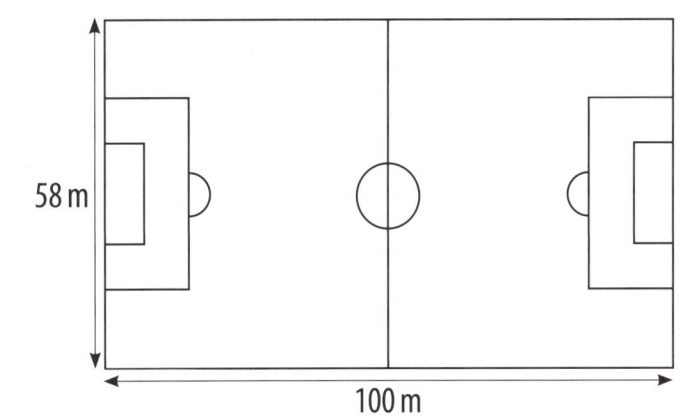

58 m
100 m

☐ 5
(1 mark)

6 The height of a model car is **7.2 centimetres**.
The length of the real car is **60 times** the height of the model.
What is the **length** of the real car?
Give your answer in **metres**. ☐ m

☐ 6
(2 marks)

Top tip
* Check the unit of measurement in each question and answer.

☐ /7

Total for this page

33

MEASUREMENT

Mass

To achieve the expected standard, you need to:
★ use, read, write and convert between units of **mass**
★ use all four operations to solve problems involving mass
★ use decimals to two decimal places.

1 What is the total mass of these objects? ☐ g

2.5 kg   560 g   1.25 kg

(1 mark)

2 A recipe uses the mass of potatoes shown on this scale.

What mass of potatoes will be left in a 2.5 kg bag? ☐

(2 marks)

3 All three objects have the same mass.

What is the mass of **one** of the objects? ☐ kg

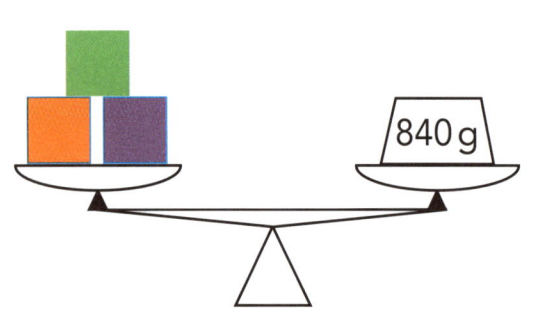

840 g

(1 mark)

4 A box of 12 tins has a mass of 5.5 kg.

The box without the tins has a mass of 520 g.

What is the mass of each tin? ☐ g

(2 marks)

Top tip
- Writing down the calculations for each step can help organise your thinking, catch errors and may even gain you a mark.

/ 6

Total for this page

Capacity

To achieve the expected standard, you need to:
★ use, read, write and convert between metric units of **capacity**
★ use all four operations to solve problems involving capacity
★ use decimals to two decimal places.

1 How many millilitres are in 4.65 litres? ☐ ml

(1 mark)

2 3 litres = 1.75 litres + ☐ ml + ☐ ml

(1 mark)

3 How many 250 ml cups can be filled from the water in this jug? ☐

(1 mark)

4 A car uses 0.2 litres of petrol to travel a distance of 6 km. How far will it travel on 1 litre of petrol? ☐ km

(1 mark)

5 In a race, teams fill buckets of water that hold 1.25 litres.

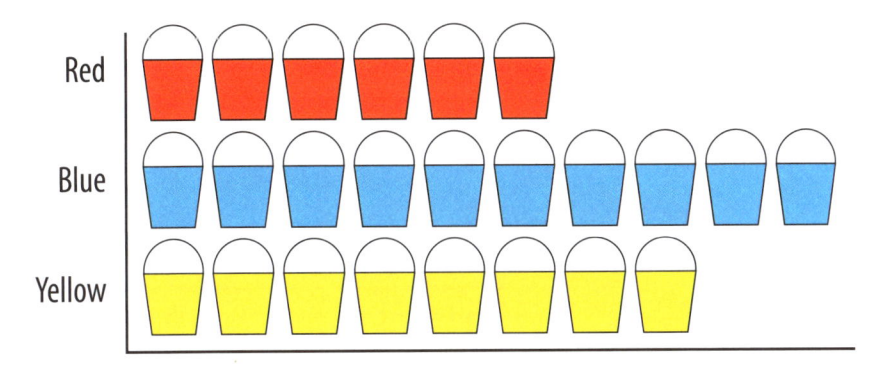

Red
Blue
Yellow

a) Write the total amount of water collected by the Yellow team in **millilitres**. ☐ ml

(1 mark)

b) How much **more** water did the Blue team collect than the Red team? ☐ l

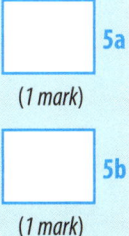

(1 mark)

Top tip
- Check the unit of measurement required each time. If you give an answer in millilitres when the question asks for litres, you will lose marks.

/ 6

Total for this page

MEASUREMENT

Money

To achieve the expected standard, you need to:
★ use all four operations to solve problems involving money, using decimal places.

1 a) What is the cost of **five** packs of pencils? ☐

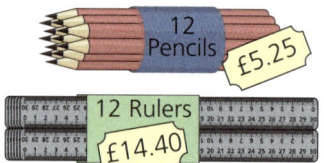

   12 Pencils £5.25
   12 Rulers £14.40

   b) What is the cost of **one** ruler? ☐

   ☐ 1a (1 mark)
   ☐ 1b (1 mark)

2 Sami buys two of these items.

   He gets £1.76 change from a £20 note.

   What are the prices of the **two** items he bought? ☐ and ☐

   £6.99 AFC
   £8.99
   £11.25
   £10.25

   ☐ 2 (2 marks)

3 Kelsey and Toby have saved £7.65 altogether.

   Kelsey has saved £1.25 **more** than Toby.

   How much did each child save?

   Kelsey ☐   Toby ☐

   ☐ 3 (2 marks)

4 The bar chart shows how much money Tom spent in different ways last month.

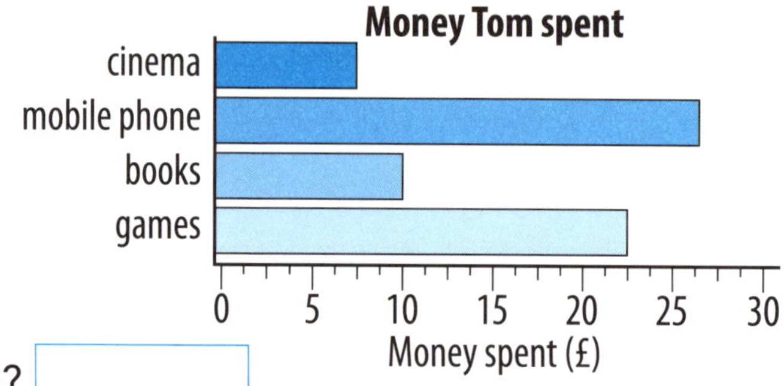

   Money Tom spent
   cinema, mobile phone, books, games
   Money spent (£)

   a) How much **more** did he spend on his mobile phone than at the cinema? ☐

   b) How much did Tom spend in **total** last month? ☐

   ☐ 4a (1 mark)
   ☐ 4b (1 mark)

5 David, Yannis and Marie share £150 prize money.

   David's share is $\frac{2}{5}$ of the total prize money and Marie's share is $\frac{1}{4}$ of the total prize money.

   How much money does Yannis get? ☐

   ☐ 5 (2 marks)

Top tip
• Always check you have answered the question that has been asked.

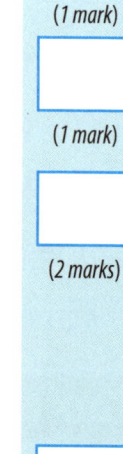

/ 10

Total for this page

MEASUREMENT

Time

To achieve the expected standard, you need to:
★ read, write and convert time between analogue (including clock faces using Roman numerals) and digital 12- and 24-hour clocks
★ use a.m. and p.m. where necessary.

1 Draw lines to match these times.

noon   25 to 8 in the evening

17:35   12 p.m.

19:25   7:25 p.m.

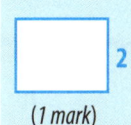

[clock showing ~7:25] p.m.   5:35 p.m.

(2 marks)

2 Order the following times from **earliest** to **latest**. Write the letters of each clock in the boxes.

A [analogue clock with Roman numerals, ~9:15] p.m.   B [21:35]   C [analogue clock ~6:00] a.m.   D [12:00]

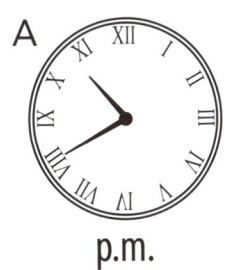

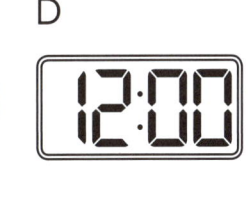

earliest ☐ ☐ ☐ ☐ latest

(1 mark)

3 Write the time that is 75 minutes **earlier** than the one shown here.

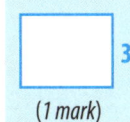

[clock showing ~2:45] ☐

(1 mark)

4 Write the time that is **halfway** between 10:20 a.m. and 13:10

☐

(1 mark)

Top tip
• Think about the number of minutes in an hour.

/ 5

Total for this page

MEASUREMENT

Time problems

To achieve the expected standard, you need to:
★ calculate the duration of an event using appropriate units of time.

1 A film starts at 9:40 p.m. and ends at 11:25 p.m.
How long does it last? ☐

(1 mark) **1**

2 Pete's art lesson ends at 14:45
It lasts for 55 minutes.
What time does it start? ☐

(1 mark) **2**

3 The shaded part of the scale shows the length of time that Mr Brown parked his car in the car park.

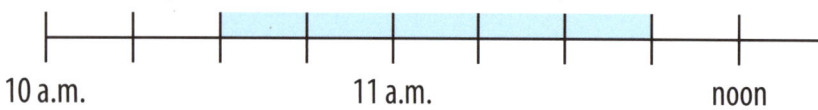

|—|—|—|—|—|—|—|—|—|
10 a.m.   11 a.m.   noon

a) What time did he arrive at the car park? ☐

(1 mark) **3a**

b) The cost of parking a car here is 35p for 15 minutes.
How much did it cost Mr Brown to park his car? ☐

(1 mark) **3b**

4 This timetable shows the times that buses stop at different places in the town.

Bus garage	10:18	11:18	12:18
Post Office	10:46	11:46	12:46
Station	11:15	12:15	13:15
School	12:08	13:08	14:08

a) How long does it take for the bus to travel from the **bus garage** to the **school**? ☐

(1 mark) **4a**

b) Mrs Lang arrives at the Post Office bus stop at 11:55
How long must she wait for the next bus? ☐

(1 mark) **4b**

c) Jasper wants to get the 13:15 bus from the station.
He gets to the station bus stop 18 minutes early.
What time does he get to the station bus stop? ☐

(1 mark) **4c**

Top tip
- Remember to check any scales first.

/7

Total for this page

38

Perimeter

To achieve the expected standard, you need to:
★ calculate the **perimeter** of **compound shapes**, when all side lengths are known or can be easily determined.

1 Calculate the perimeter of this square. ☐ cm

(1 mark)

2 Calculate the perimeter of this shape. ☐ cm

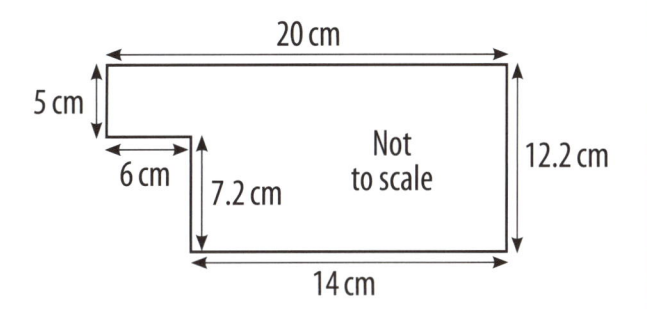

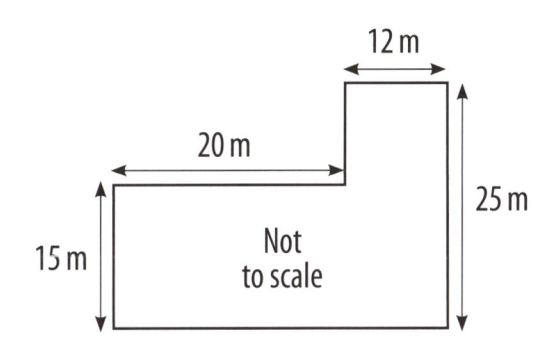

(1 mark)

3 Calculate the perimeter of this shape. ☐ m

(1 mark)

4 A square and a rectangle are placed together to make this new shape.

Calculate the perimeter of the new shape.

Write your answer in **centimetres**.

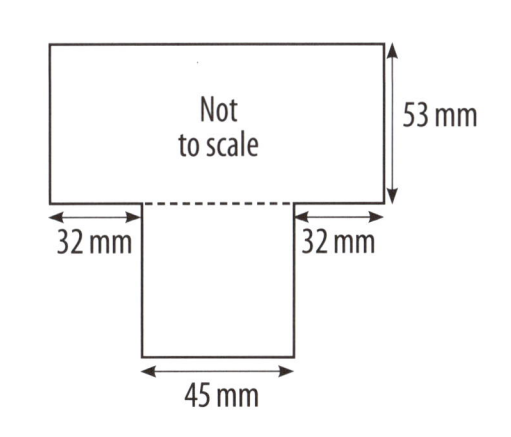

☐

(2 marks)

Top tip
- Look for any given measurements. Think how you can use them to work out unknown measurements.

/5

Total for this page

MEASUREMENT

Estimating the area of irregular shapes

To achieve the expected standard, you need to:
★ estimate the **area** of irregular shapes by counting squares.

1 Draw a shape on the grid that has **half** the area of this shape.

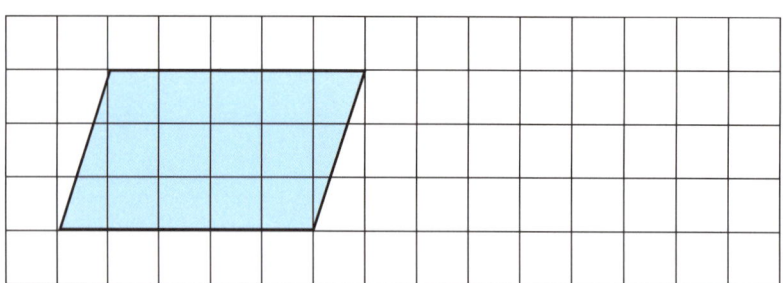

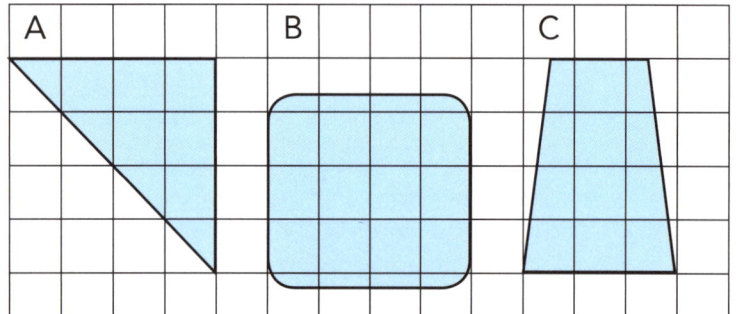

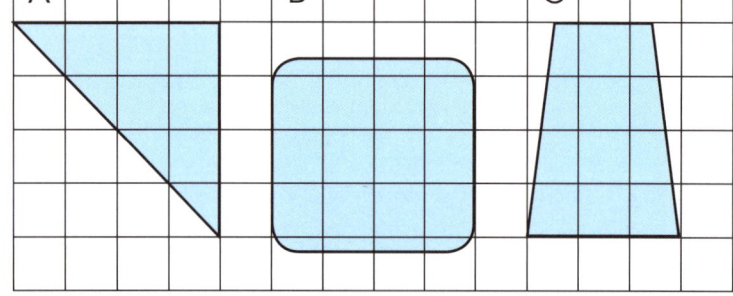

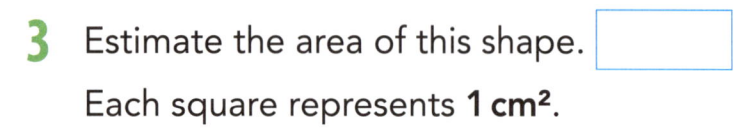

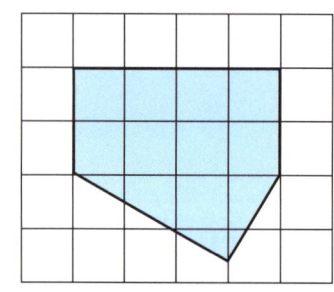

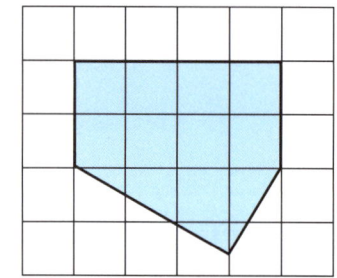

(1 mark)

2 Tick (✓) the shape with the **largest** area.

(1 mark)

3 Estimate the area of this shape. ☐

Each square represents **1 cm²**.

(1 mark)

4 Asha thinks that the area of this shape is approximately 15 m².

Each square represents **1 m²**.

Do you agree with Asha?
Circle your answer. YES / NO

Explain your answer.

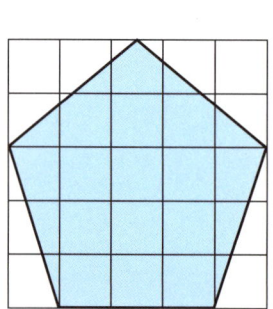

(1 mark)

Top tip
- Look for fractions of squares that can be put together to make (or nearly make) a whole square.

/ 4

Total for this page

Area by formula

MEASUREMENT

To achieve the expected standard, you need to:
★ calculate and compare the area of squares and other rectangles
★ use standard units: square centimetres (cm²) and square metres (m²).

1 A square has sides of 11 cm. Calculate its area. ☐ cm²

☐ 1
(1 mark)

2 Which of these shapes has the **largest area**? ☐

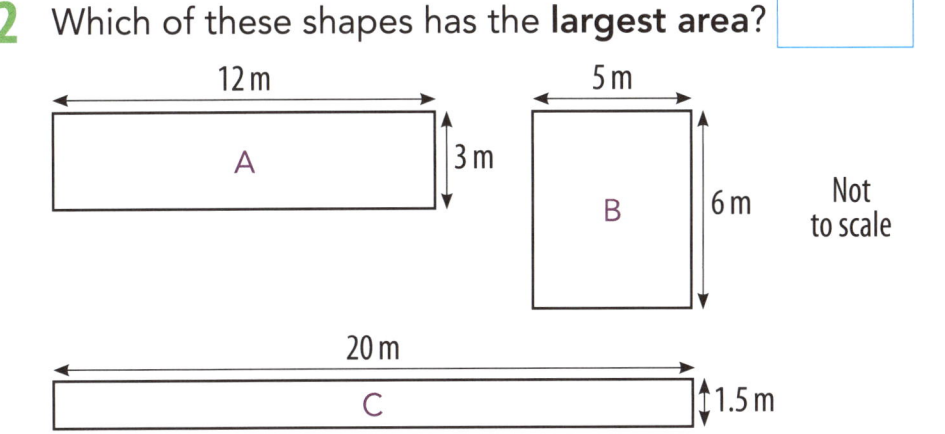

- A: 12 m × 3 m
- B: 5 m × 6 m
- C: 20 m × 1.5 m

Not to scale

☐ 2
(1 mark)

3 Jamie draws a rectangle with an area of **36 cm²**.

What are the lengths of its sides?

Find two possible answers.

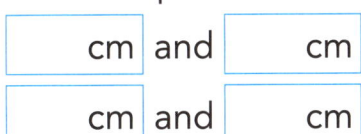

☐ cm and ☐ cm
☐ cm and ☐ cm

☐ 3
(1 mark)

4 How much **larger** is the area of the rectangle than the area of the square? ☐ m²

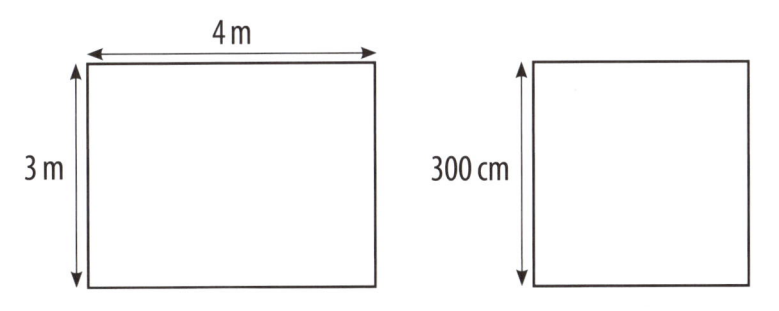

- Rectangle: 4 m × 3 m
- Square: 300 cm

☐ 4
(2 marks)

Top tip
- Think about how multiplication is related to area.

/ 5
Total for this page

41

GEOMETRY – PROPERTIES OF SHAPES

Drawing lines and angles

To achieve the expected standard, you need to:
★ complete simple 2-D shapes using given lengths and **acute angles** that are multiples of 5 **degrees**.

1 Draw a rectangle with lengths 6.5 cm and 3.5 cm.

☐ 1
(1 mark)

2 A right-angled triangle has a base of 11 cm.

a) What is the height of the triangle (x)? ☐ cm

b) Measure and label the sizes of each angle.

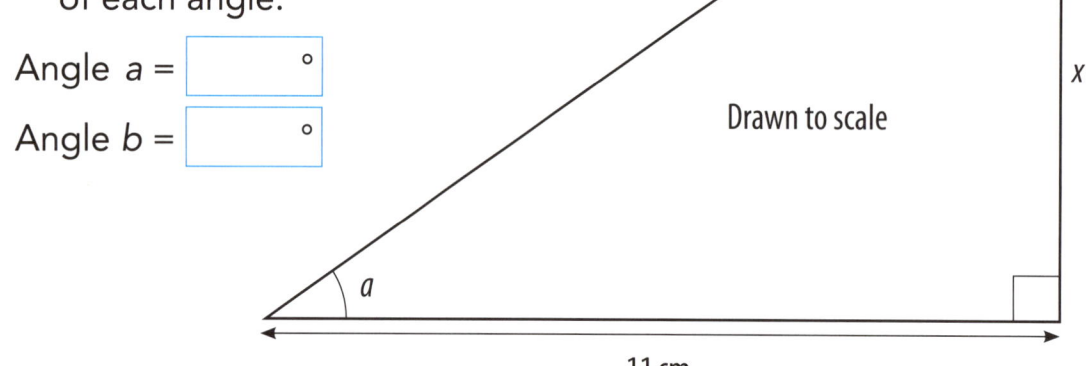

Angle a = ☐ °

Angle b = ☐ °

Drawn to scale

11 cm

☐ 2a
(1 mark)

☐ 2b
(1 mark)

3 a) Complete the **trapezium** with angle a = 65°.

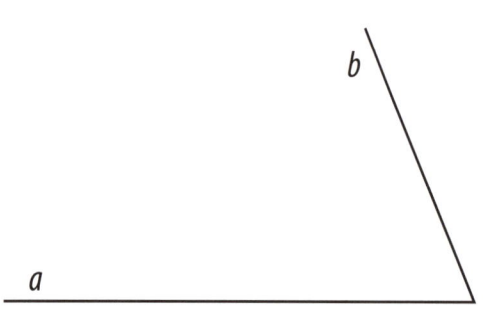

b) Write the size of angle b to the **nearest 5 degrees**. ☐ °

☐ 3a
(1 mark)

☐ 3b
(1 mark)

Top tip
- Think how you can be sure that an angle is a right angle.
- Remember that lines or angles on shapes **drawn to scale** can be **measured**.

☐ / 5

Total for this page

GEOMETRY – PROPERTIES OF SHAPES

2-D shapes

To achieve the expected standard, you need to:
★ recognise properties of 2-D shapes in terms of number and length of sides, **parallel** sides, type and size of angles
★ identify **irregular** and **regular** polygons
★ name 2-D shapes according to their properties.

1 Complete the table.

Name of 2-D shape	Number of straight sides	Number of right angles
rectangle		
	3	1
regular hexagon		
equilateral triangle		

(1 mark)

2 Tick (✓) all the **hexagons**.

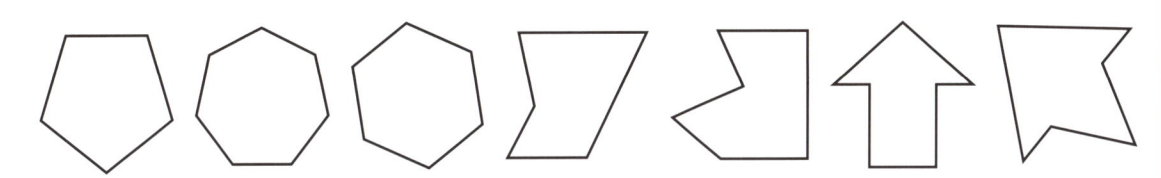

(1 mark)

3 Anya wants to draw a shape with **at least** two pairs of **parallel** sides.

Circle **all** the shapes she can draw.

trapezium   rectangle   rhombus   pentagon   regular octagon

(1 mark)

4 Decide which shape is being described by the table.

Complete the table.

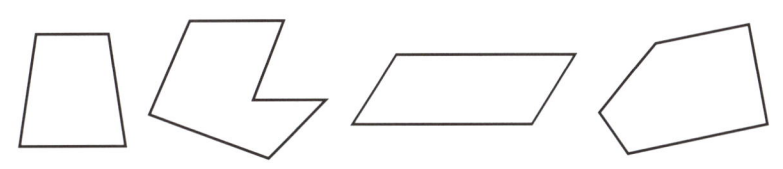

Name of shape	Number of pairs of parallel lines	Number of pairs of perpendicular lines	Total number of sides
	1	2	

(1 mark)

Top tip

• Use the corner of a ruler to help identify any acute angles.

/ 4

Total for this page

43

GEOMETRY – PROPERTIES OF SHAPES

3-D shapes

To achieve the expected standard, you need to:
★ recognise properties of 3-D shapes – **faces**, **edges** and **vertices**
★ recognise and describe simple 3-D shapes
★ use **nets** and other 2-D representations of simple 3-D shapes.

1 Fill in the missing information about this 3-D shape.

Number of faces ☐
Number of edges ☐
Number of vertices ☐

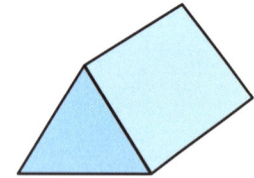

☐ 1
(1 mark)

2 Write the letters of the shapes in the correct place in the Venn diagram.

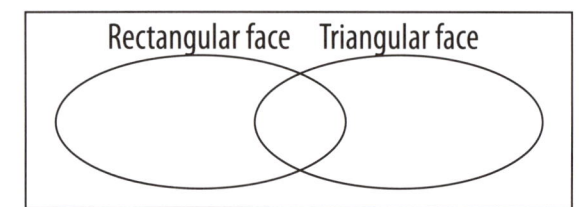

Rectangular face   Triangular face

A cone
B triangular prism
C tetrahedron
D cuboid
E sphere

☐ 2
(1 mark)

3 What solid shape is Emma describing?

The solid shape is a _____

My solid shape has five faces and eight edges. Four faces are the same shape but one face is square.

☐ 3
(1 mark)

4 Tick (✓) the nets that fold to make a **cube**.

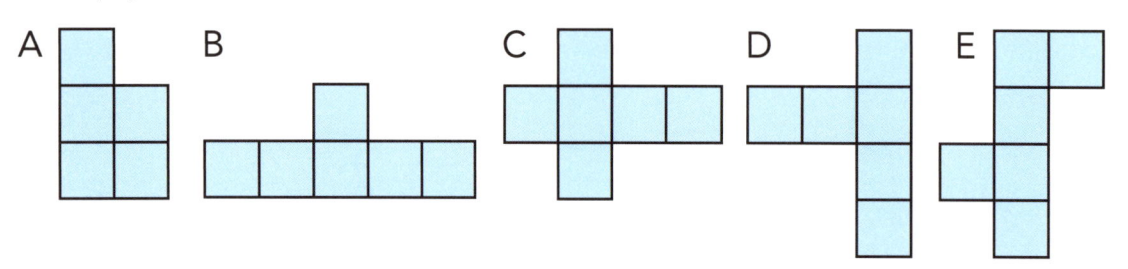

A   B   C   D   E

☐ 4
(1 mark)

5 Tick (✓) the nets that fold to make a **triangular prism**.

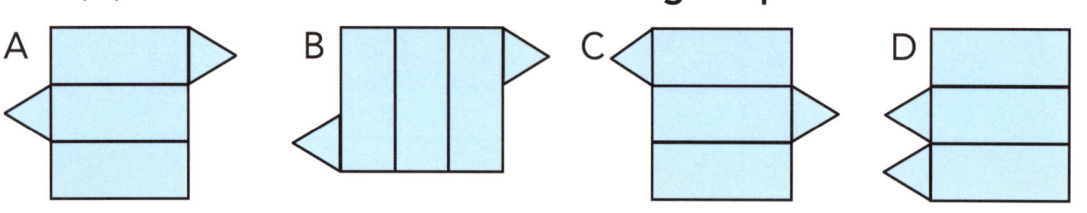

A   B   C   D

☐ 5
(1 mark)

Top tip
• A picture of a 3-D shape may not show all the faces, edges and vertices.

/ 5

Total for this page

GEOMETRY – PROPERTIES OF SHAPES

Angles and degrees

To achieve the expected standard, you need to:
★ recognise angles where they meet at a point or on a straight line
★ find missing angles at a point and along a straight line.

1 Find the size of angle m. ☐°   Not to scale

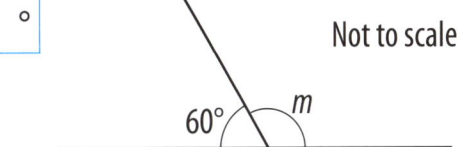

(1 mark)

2 Find the size of angle a. ☐°   Not to scale

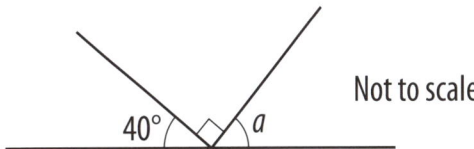

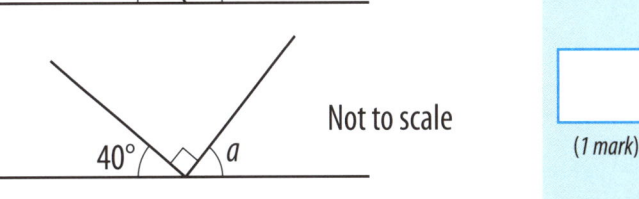

(1 mark)

3 Calculate angles a and b.

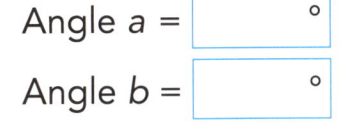

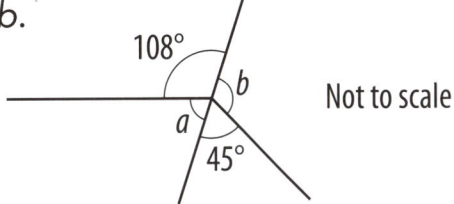

Angle a = ☐°

Angle b = ☐°

Not to scale

(1 mark)

4 Calculate the size of angle p. ☐°

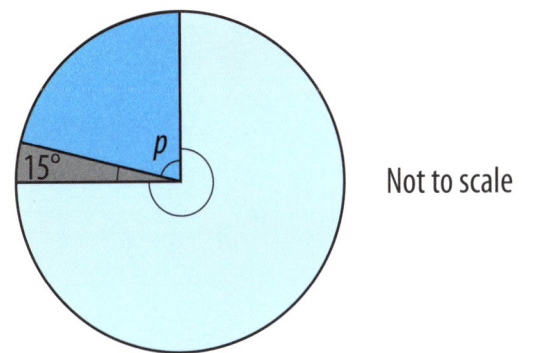

Not to scale

(1 mark)

5 Calculate the size of angle q. ☐°

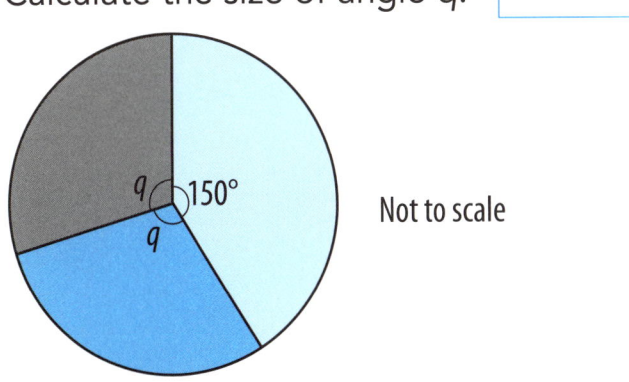

Not to scale

(2 marks)

Top tip
• Look out for the symbol that represents a right angle in missing angle questions.

/ 6

Total for this page

45

GEOMETRY – PROPERTIES OF SHAPES

Angles in triangles

To achieve the expected standard, you need to:
★ find unknown angles in different types of triangles.

1. Calculate the size of the angle m. ☐°

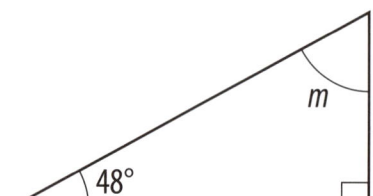

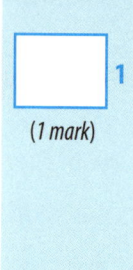

   (1 mark)

2. Write the size of **each** angle on the triangle.

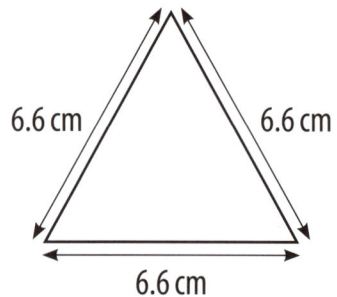

   6.6 cm, 6.6 cm, 6.6 cm

   (1 mark)

3. Work out the size of angles a and b.
   Angle a = ☐°
   Angle b = ☐°

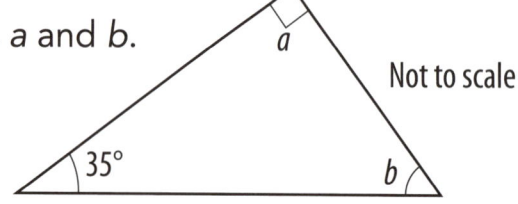

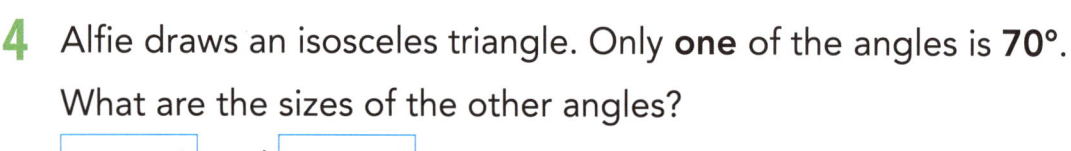

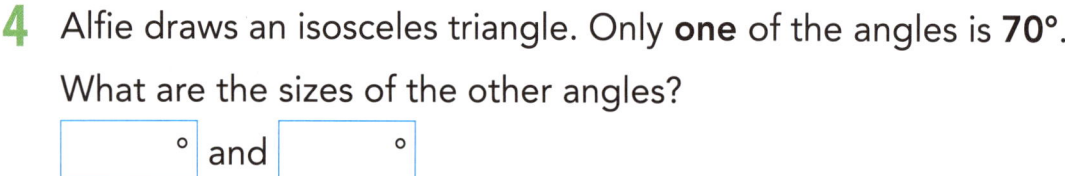

   Not to scale
   35°

   (1 mark)

4. Alfie draws an isosceles triangle. Only **one** of the angles is **70°**.
   What are the sizes of the other angles?
   ☐° and ☐°

   (1 mark)

5. Calculate the sizes of angles a and b.
   a = ☐°
   b = ☐°

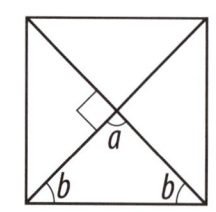

   (1 mark)

Top tip

- Remember that an isosceles triangle has two angles the same size, as well as two sides of the same length.

/ 5

Total for this page

GEOMETRY – POSITION AND DIRECTION

Coordinates

To achieve the expected standard, you need to:
★ describe the positions on a 2-D coordinate grid using axes with equal scales in the first quadrant and become more confident in plotting points in all four quadrants
★ use **coordinates** to complete a given rectangle.

1 Write the coordinates of each vertex of the triangle.

A (,)
B (,)
C (,)

(1 mark)

2 Write the coordinate of vertex **D** that completes the rectangle.

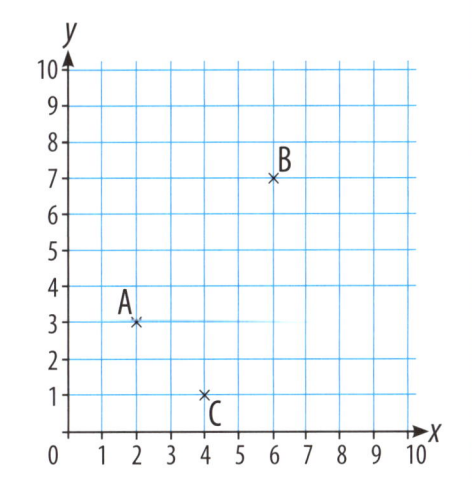

D (,)

(1 mark)

3 Plot these coordinates on the grid.

A (1,0) B (–2,2) C (–2,7) D (2,7) E (4,4)

Join the coordinates in order to make a closed shape.

Name the shape:

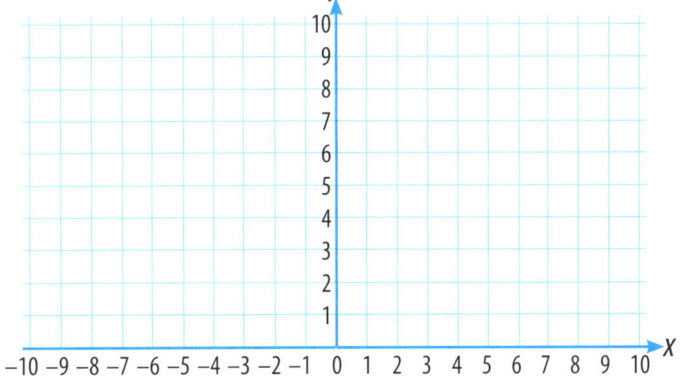

_____

(2 marks)

4 Jayden plots (5,4) and (8,4) as two vertices of a square.

What could the coordinates of the other corners be?

(,) and (,)

Top tip
- Coordinates are always written in the order (x,y).

(1 mark)

/ 5

Total for this page

GEOMETRY – POSITION AND DIRECTION

Translations

To achieve the expected standard, you need to:
★ identify, describe and represent the position of a shape following a **translation**.

1 Write the **new** position of coordinate M after the translation **2 squares left** and **3 squares down**.

M (,)

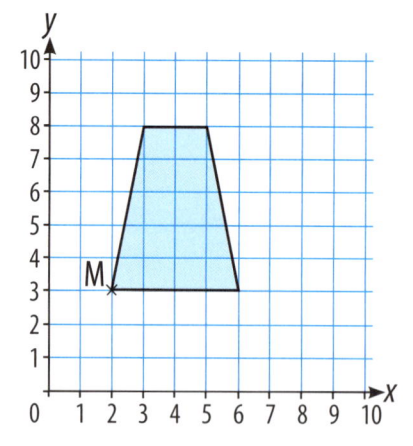

(1 mark)

2 Draw the position of the square after the translation **3 squares right** and **1 square up**.

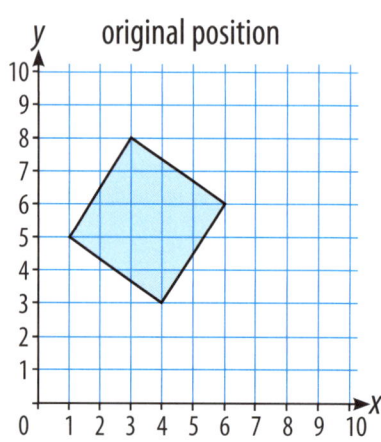

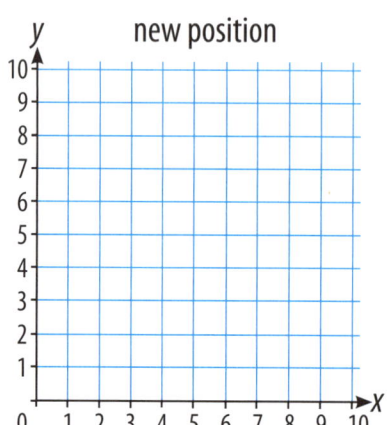

(1 mark)

3 Describe the translation of the triangle.

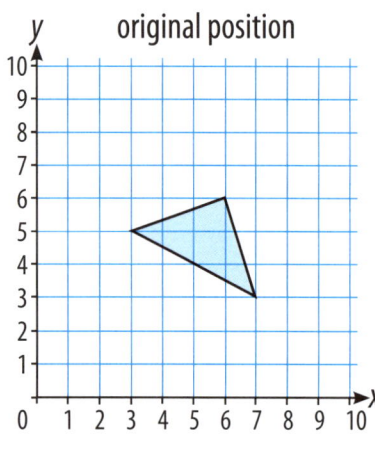

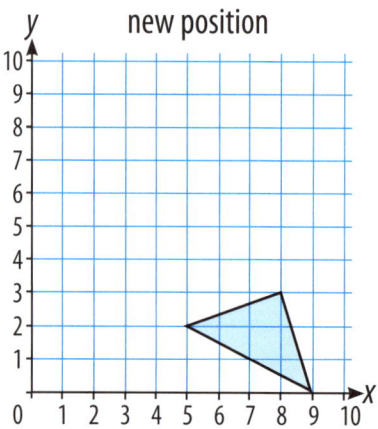

_____

_____

(1 mark)

Top tip
- Imagine a translation as a shape sliding horizontally and vertically. The size and orientation of the shape do not change, just the position.

/3

Total for this page

GEOMETRY – POSITION AND DIRECTION

Reflections

To achieve the expected standard, you need to:
★ identify, describe and represent the position of a shape following a **reflection**.

1 Tick (✓) the shape that has been **correctly** reflected in the mirror line.

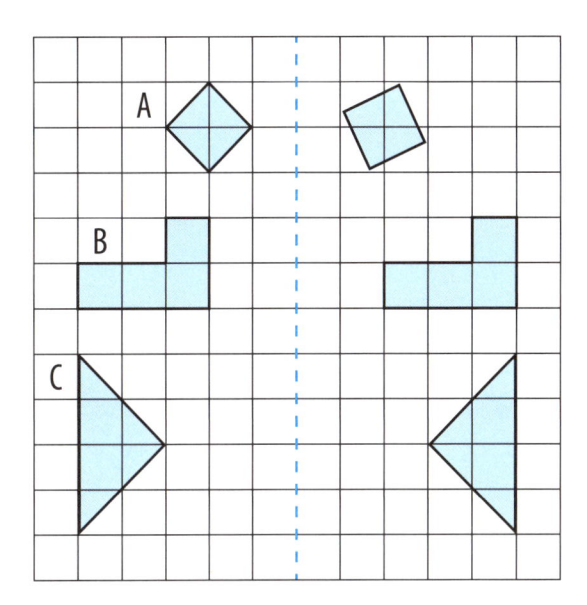

2 Reflect the shape in the mirror line.
Draw the reflected shape.

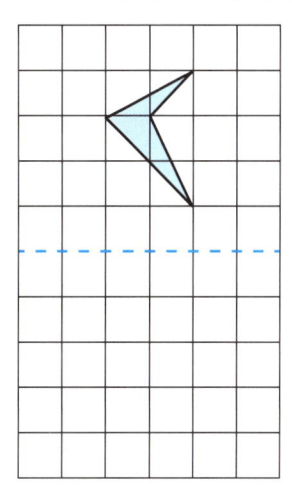

(1 mark)

(1 mark)

3 Reflect the shape in the mirror line.
Draw the reflected shape.

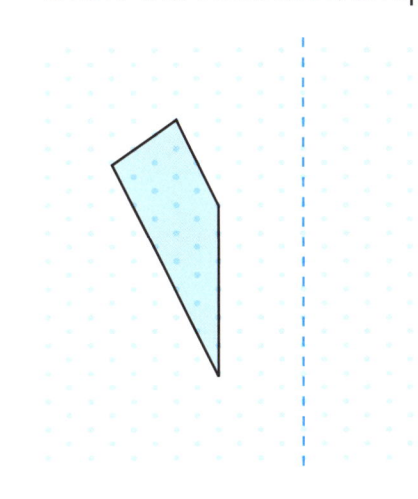

4 Reflect the shape in the **x axis**.
Draw the reflected shape.

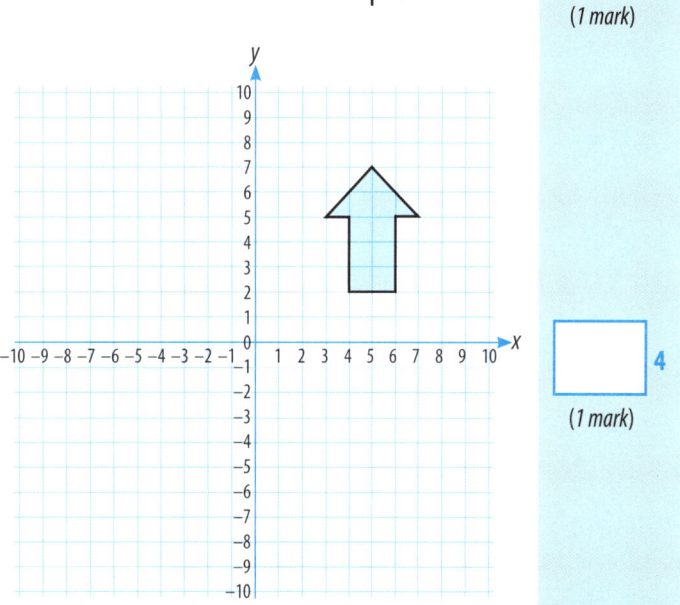

(1 mark)

(1 mark)

Top tip
- Shapes do not change size when they are reflected.

/ 4

Total for this page

49

GEOMETRY – POSITION AND DIRECTION

Reflective symmetry

To achieve the expected standard, you need to:
★ compare and classify 2-D shapes in terms of **reflective symmetry**.

1 Tick (✓) the shapes that have reflective symmetry.

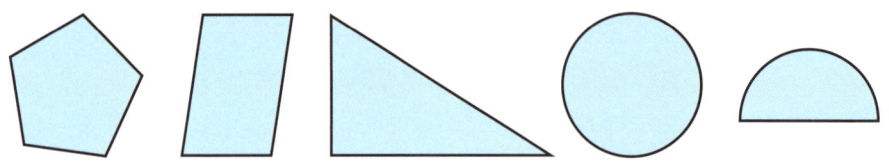

(1 mark)

2 Draw the line of symmetry on the shape.

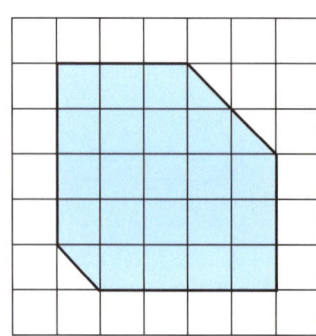

(1 mark)

3 Write the letters of the shapes in the correct place in the Carroll diagram.

	Reflective symmetry	No reflective symmetry
Quadrilateral		
Not quadrilateral		

(1 mark)

4 Shade **one more** square to make this shape symmetrical.

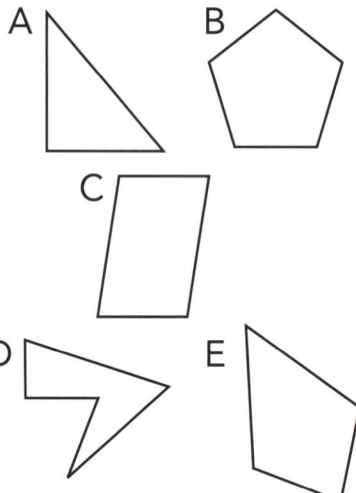

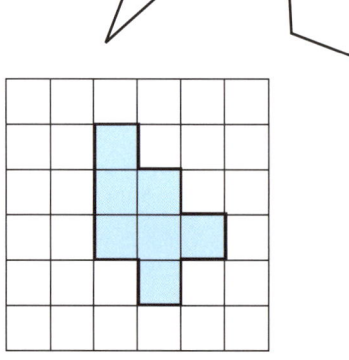

(1 mark)

Top tip
- Use a real mirror to help you. Draw mirror lines using a ruler.
- It is often useful to turn the paper so that you can look at a shape in different ways.

/ 4

Total for this page

Tables

To achieve the expected standard, you need to:
★ complete, read and interpret information in **tables**.

1 Complete the table showing the children's favourite sport.

Favourite sport	Tally	Total
badminton	|||| |||| |||| |||| |||	
hockey	|||| |||| ||	12
football	|||| |||| |||| |||	
		53

(1 mark) **1**

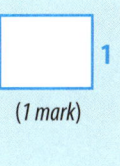

2 a) How much money have the children saved in total? ☐

	Sally	Ash	Peta	Pav
Money saved	£7.75	£10.99	£9.20	£11.49

(1 mark) **2a**

b) How much **more** has Pav saved than Sally? ☐

(1 mark) **2b**

3 The table shows the days of the week the players are free to play a tennis match.

	Monday	Tuesday	Wednesday	Thursday	Friday
Kate	✓		✓		✓
Zack		✓			✓
Jayden		✓		✓	
Dina	✓			✓	✓

a) On which day of the week can **no** matches be played?
_____

(1 mark) **3a**

b) Which two players are **not** able to have a match together?
_____ and _____

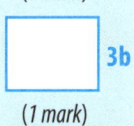

(1 mark) **3b**

4 This table shows how much water was collected in a week in three different containers.

	Container A	Container B	Container C
Amount of water container holds	$3\frac{1}{2}$ litres	$4\frac{1}{4}$ litres	$2\frac{3}{4}$ litres
Number of containers filled	4	3	2

a) How much **more** water can container B hold than container C? ☐ litres

(1 mark) **4a**

b) How many litres have been collected **altogether**? ☐ litres

(2 marks) **4b**

/ 8

Total for this page

STATISTICS

Pictograms

To achieve the expected standard, you need to:
★ interpret and present data using **pictograms**.

1 Complete the pictogram using the information shown in the table.

	Total points
Red	12
Blue	16
Green	20

Red
Blue
Green

◉ = 3 points

(1 mark) 1

2 a) How many **more** children like canoeing than treasure hunts?

b) How many children like outdoor activities in **total**?

c) Nine children do **not** like any outdoor activities.

Show this on the pictogram using the same symbols.

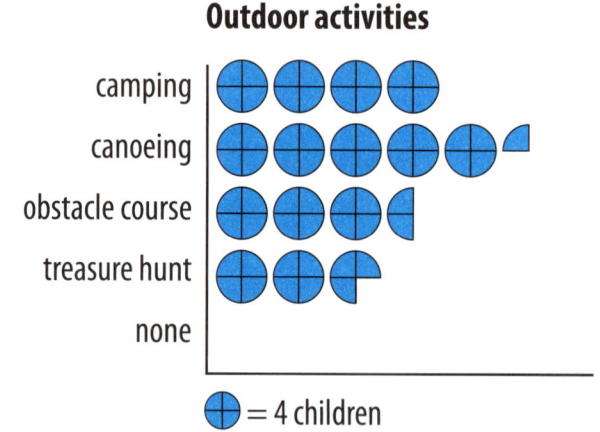

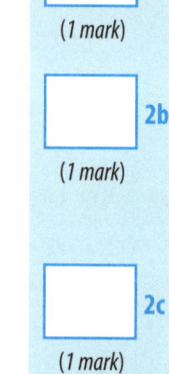

Outdoor activities

camping
canoeing
obstacle course
treasure hunt
none

● = 4 children

(1 mark) 2a
(1 mark) 2b
(1 mark) 2c

3 a) Which magazines sold **fewer** than 25 copies?

_____

_____

b) Which two magazines sold a **total** of 52 copies?

_____

and

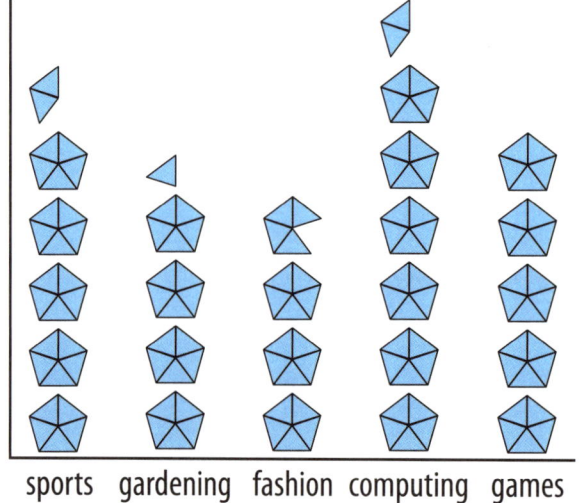

_____

Number of magazines sold in one weekend

sports gardening fashion computing games

⬟ = 5 magazines

(1 mark) 3a
(1 mark) 3b

★ **Top tip**
- Check the value of each part of the pictogram before tackling the question.

/ 6

Total for this page

52

Bar charts

STATISTICS

To achieve the expected standard, you need to:
★ complete, read and interpret information presented in **bar charts**.

1 This bar chart shows the types of library books borrowed by Class 6 last term.

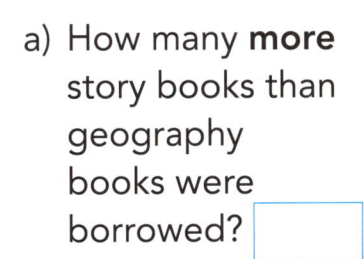

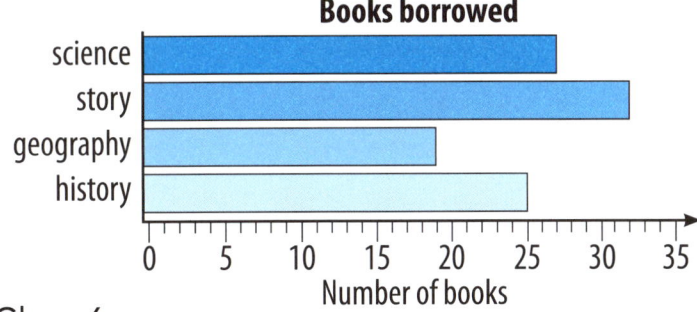

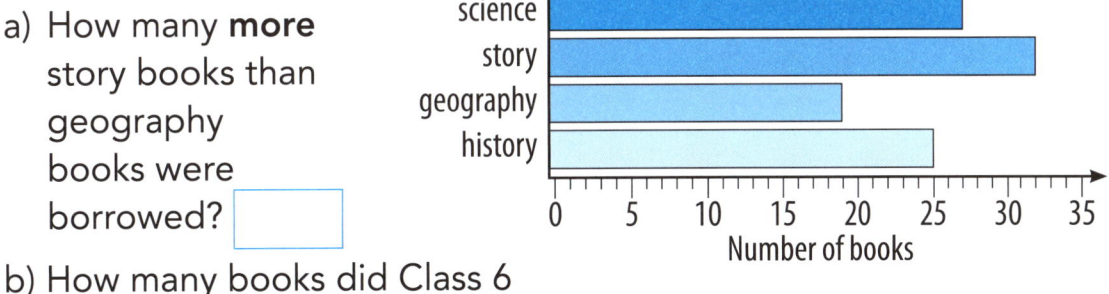

a) How many **more** story books than geography books were borrowed?

b) How many books did Class 6 borrow in **total** from the library?

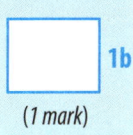

2 This bar chart shows the mass of different objects.

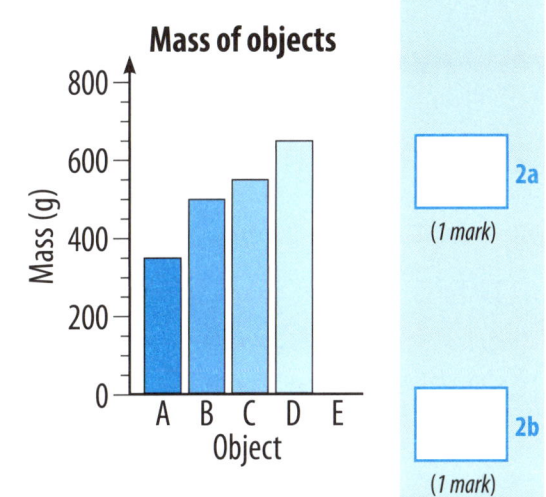

a) Complete the table using the bar chart to help you.

	Object A	Object B	Object C	Object D
Mass (g)				

b) Object E has a mass of 450 g. Draw the missing bar on the chart.

3 This bar chart shows the money raised by each class in a cake sale.

a) How many classes raised **more** than £20?

b) Cherry Class raised £1.25 **less than** Oak Class. Draw the missing bar on the chart.

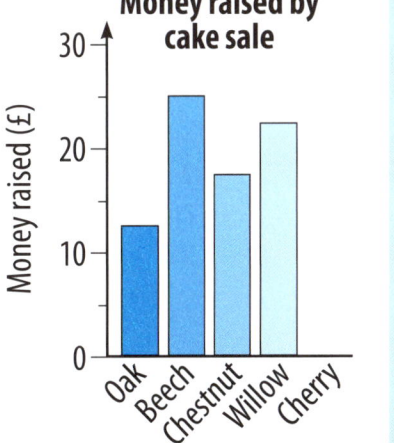

c) The school cake sale target was **£100 altogether**.

How much **more** money did they need to raise to reach the target?

> **Top tip**
> - Look at the scale to see what each interval represents.

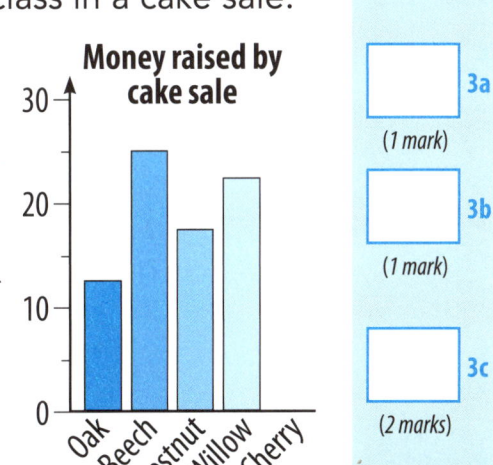

STATISTICS

Pie charts

To achieve the expected standard, you need to:
★ interpret simple **pie charts**.

1 The pie chart compares the different types of small pets owned by 120 children.

a) What **fraction** of the children do **not** have **fish**?

b) How many children do **not** have fish?

c) Ed says the total number of children with fish and gerbils is **equal** to the number of children with hamsters. Is he correct? Circle your answer. YES / NO

Explain your answer.

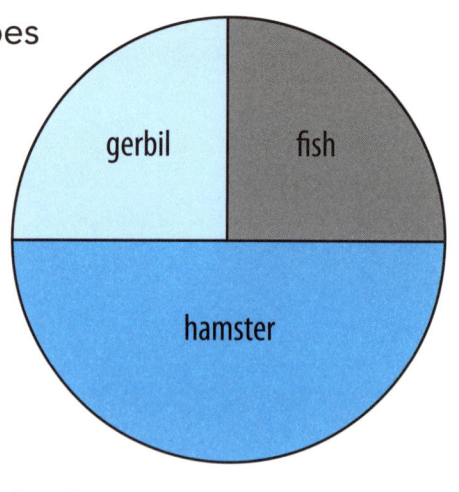

_____

_____

2 This pie chart compares the money spent on different items for a classroom. £240 is spent in total.

a) What **fraction** of the total amount is spent on pens?

b) How **much money** is spent on **other items**?

c) What is the **difference** between the money spent on pens and the money spent on pencils?

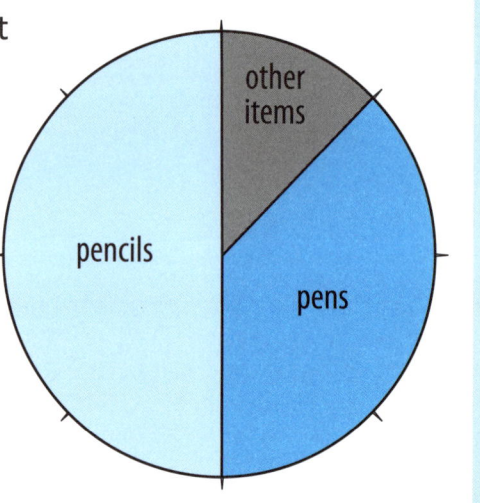

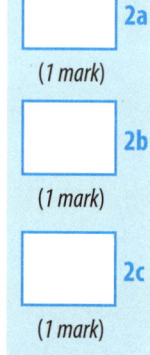

Top tip
- Use what you know about fractions of a circle to help with pie chart questions.

Line graphs

To achieve the expected standard, you need to:
★ interpret **line graphs** to solve problems.

1 The graph shows the growth of a bean plant.

a) Use the graph to help you complete the table.

Day	4	8	14	20
Height (mm)				

b) Estimate the height of the bean plant on **day 24**. ☐ mm

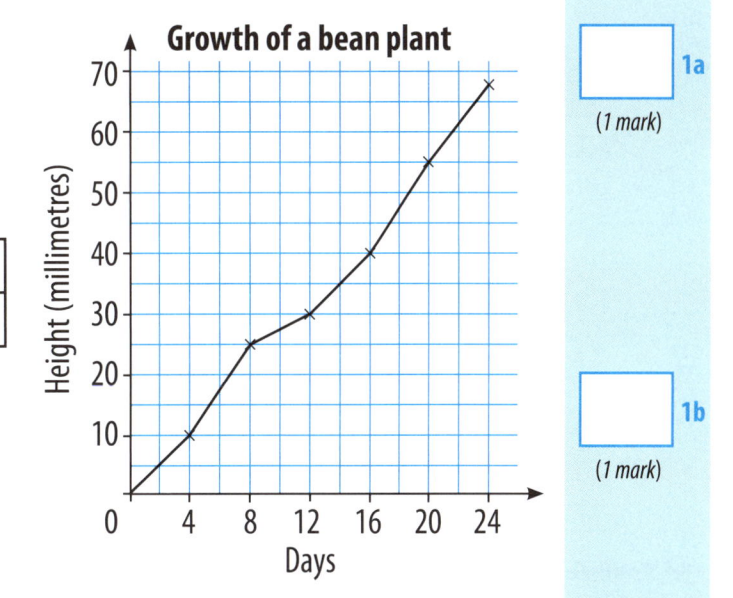

Growth of a bean plant

2 This graph shows the temperature outside at different times of the day.

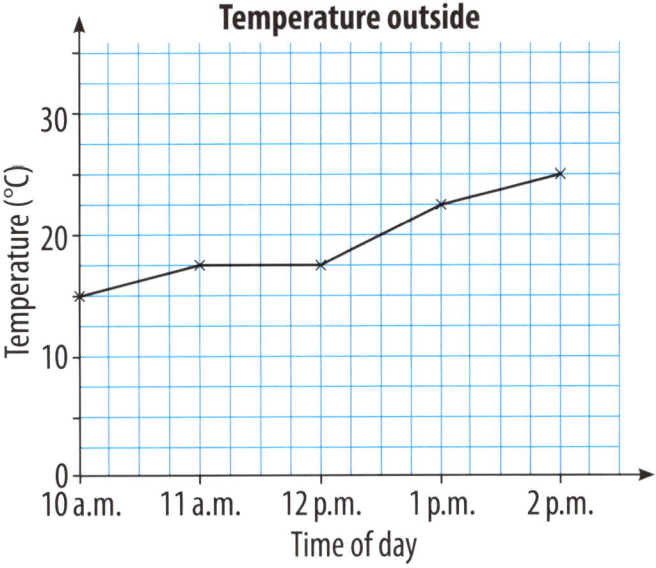

Temperature outside

a) How much warmer is the temperature at **1 p.m.** than at **11 a.m**?
☐ °C

b) What was the temperature at **12:30 p.m**? ☐ °C

c) For how long was the temperature **below 20°C**?
_____

Top tip

- Look carefully at the intervals and scales on each graph.

STATISTICS

Averages

To achieve the expected standard, you need to:
★ calculate and interpret the **mean** as an **average** for simple sets of discrete data.

1 Calculate the mean of the children's test scores. ☐

Oscar	Harriet	Lewis	Katy	Theo
9	7	11	12	6

(1 mark)

2 Calculate the mean number of buckets of water collected by teams in a race. ☐

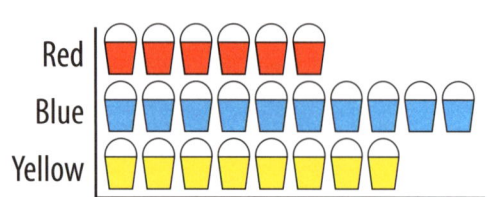

Red, Blue, Yellow

(1 mark)

3 Find the mean of these capacities.

$1\frac{1}{2}$ litres   3 litres $3\frac{1}{2}$ litres   2 litres   ☐ litres

(1 mark)

4 15 11 ? The **mean** of these three numbers is **12**
What is the missing number? ☐

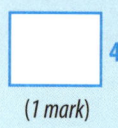

(1 mark)

5 Find the mean of these lengths.

90 cm   60 cm   0.5 m   30 cm   ☐ cm

(1 mark)

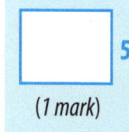

6 Calculate the average number of **minutes** per weekday that children spend playing sports. ☐ minutes

Monday	Tuesday	Wednesday	Thursday	Friday
30 minutes	1.5 hours	45 minutes	2 hours	35 minutes

(2 marks)

Top tip
- Always check the unit of measurement, and convert any that are in a different unit.

/7

Total for this page

Answers

Number and place value

Place value of whole numbers (page 6)
1. 9 hundred (900), 9 thousand (9,000), 9 tens (90), 9 hundred thousand (900,000)
2. Award **2 marks** for all four correct
   Award **1 mark** for any three correct

   7 thousand — 1,233,420
   900 thousand — 597,302
   30 thousand — 3,095,799
   3 million — 979,200

   (7 thousand → 3,095,799; 900 thousand → 979,200; 30 thousand → 1,233,420; 3 million → 597,302 — shown by crossing lines)
3. Any two numbers that use all the digit cards to make two numbers between 5,750 and 6,250 (e.g. 5,780 and 6,134)
4. 3,463,400 and 3,483,400
5. 9,000

Comparing and ordering whole numbers (page 7)
1. a) 34,601 > 34,599
   b) 709,898 < 709,988
   Award **1 mark** for both answers correct.
2. 940,500 g   904,500 g   95,400 g   94,500 g   9,500 g
3. e.g. 2,5**1**0,500   2,519,**5**85   2,**6**05,250   2,60**6**,125
4. All years must be ordered correctly.

	Year
Greatest number of visitors	2015
	2013
	2012
	2016
	2017
Fewest number of visitors	2014

Rounding (page 8)
1. 4 for ☐78 and 0;1;2;3; or 4 for 5☐0
2. a) 130   b) 100   c) 540
   Award **1 mark** for all answers correct.
3. 952; 1,045; 1,009
4. Award **2 marks** for all correct.
   Award **1 mark** for at least six correct.

	To the nearest 10	To the nearest 1,000	To the nearest 10,000
45,419	45,420	45,000	50,000
234,549	234,550	235,000	230,000
3,126,095	3,126,100	3,126,000	3,130,000

5. a) 5,745
   b) 5,749

Place value of decimal numbers (page 9)
1. 10
2. 1.93; 1.83; 1.78
3. 2.59; 2.68; 2.77; 2.86; 2.95
4. 0.048
5. Ben is incorrect because 6 equals 6 ones and 0.06 equals 6 hundredths, which is less than 6 ones.
6.

0.72	2.70	7.02	7.2	27.0

   smallest .... largest

Negative numbers (page 10)
1. 1; −5
2. 5 levels
3. −7
4. a) Two different pairs, e.g. 10 and −1, 9 and −2, 8 and −3, 7 and −4, 6 and −5, 5 and −6
   b) Award **1 mark** for a response of NO, and an explanation to show that either:
   - 11 and 0 have a difference of 11 but 0 is not a negative number.
   - one of the numbers has to be negative and 0 is not negative (implied that 0 and 11 have the required difference).

   Do not award any marks for only identifying 11 and 0.

Number – Addition, subtraction, multiplication and division

Addition (page 11)
1. 800
2. 121,483
3. a) 15,476
   b) 22,074
4.
   ```
     3 9 4 8
   + 2 1 4 5
   ─────────
     6 0 9 3
   ```
5. Award **2 marks** for correct answer £550,696.
   Award **1 mark** for evidence of appropriate working but with an incorrect answer.
   Do not award any marks if a final answer is missing.

Subtraction (page 12)
1. 10,924
2. 160,000
3. 29,138
4. Award **2 marks** for the correct answer of £5,122.
   Award **1 mark** for evidence of the correct method with no more than one mathematical error, e.g.
   £2,759 + £5,764 + £1,905 = £10,428
   £15,550 − £10,428 = wrong answer
5.
   ```
     4 0 8 8 6
   − 2 4 6 3 7
   ───────────
     1 6 2 4 9
   ```

57

ANSWERS

Multiplying and dividing by 10 and 100 (page 13)

1. a) 23,500   b) 100   c) 0.99
   Award **1 mark** for all answers correct.
2. 456 × 100 → 45,600
   45.6 ÷ 10 → 4.56
   0.45 × 100 → 45
   4.56 ÷ 10 → 0.456
3. Award **2 marks** for all three correct.
   Award **1 mark** for any two correct.

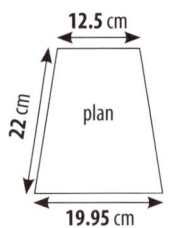

   12.5 cm
   22 cm
   plan
   19.95 cm

4. Parcel A is 25 kg and Parcel B is 2.5 kg.

Multiples and factors (page 14)

1. 77; 220; 1,100
2. 1 or 3
3. Award **2 marks** for all lines matched correctly and 3 or 6 written in the missing factor box.
   Award **1 mark** for all lines matched correctly but no or incorrect factor given or **1 mark** for factor correctly identified and one multiple correctly matched.

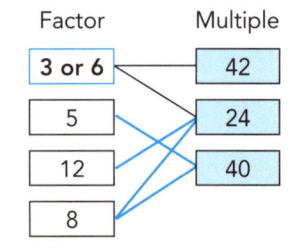

Factor	Multiple
3 or 6	42
5	24
12	40
8	

4. 1; 2; 4; 8; 16; 32
5. 2; 6; 10; 30

Multiplying by larger numbers (page 15)

1. 25,263
2. Award **2 marks** for correct answer 5,005.
   Award **1 mark** if the answer is incorrect, but there is evidence of using the formal method of long multiplication that contains no more than one arithmetic error. To gain 1 mark, working must be carried through to reach an answer. In all cases accept follow-through of one error in working, e.g.

   ```
       3 8 5
   ×     1 3
     1 1 5 5
     3 8 5 0
   wrong answer
   ```

   Do not award any marks if the final (answer) line of digits is missing or the error is in the place value, e.g. by omission of the zero when multiplying by the tens, e.g.

   ```
       3 8 5
   ×     1 3
     1 1 5 5
       3 8 5
   wrong answer
   ```

3. 8,925

4. Award **2 marks** for correct answer 3,247 × 5 = 16,235
   Award **1 mark** for correct calculation 3,247 × 5 and with worked answer containing only one arithmetic error (but not an error relating to place value).
5. Award **2 marks** for the correct answer of 14,976.
   Award **1 mark** for evidence of the correct method with no more than one mathematical error, e.g.
   52 × 24 = 1,248
   1,248 × 12 = wrong answer

Square numbers (page 16)

1. 25 and 16
2. 64p
3. e.g. 1 + 4 = 5; 4 + 9 = 13; 1 + 16 = 17; 4 + 25 = 29
4. Amy
5. Award 1 mark for a response of YES, and an explanation to show that either:
   - 81 and 100 are the only two numbers that are square between 80 and 110
   - 9^2 is 81 and 10^2 is 100, so these are the only two square numbers between 80 and 110.

Short division (page 17)

1. 62
2. 638
3. 884
4. 44
5. 477.5 or $477\frac{1}{2}$

Long division (page 18)

1. Award **2 marks** for correct answer 57.
   Award **1 mark** if answer is incorrect, but there is evidence of using the formal method of long division. Working must be carried through to reach an answer for the award of 1 mark. In all cases accept follow-through of one error in working, e.g.

   ```
              wrong answer
   15 ) 8 5 5
        7 5 ↓
        1 0 5
        1 0 5
              0
   ```

   Do not award any marks if a final answer is missing.

2. Award **2 marks** for correct answer 47 r6 or $47\frac{1}{4}$ or 47.25
   Award **1 mark** if working has been carried through to reach an answer and accept follow-through of one error (see Q1 above).
   Do not award any marks if a final answer is missing.
3. 33
4. Award **2 marks** for correct answer £65
   Award **1 mark** if the answer is incorrect, but there is evidence of an appropriate method.
5. Award **2 marks** for correct answer 21 (there is not enough ribbon to make 22).
   Award **1 mark** if the answer is incorrect, but there is evidence of an appropriate method.

Prime numbers (page 19)

1. 2; 7; 11; and 17
2. 3

… ANSWERS

3 2 and 17

4 2; 3; or 5

5 2 or 3 or 7

6 2 + 11 + 13 **and** 2 + 7 + 17 **and** 2 + 5 + 19

Award **2 marks** for **three sets** of correct responses.

Award **1 mark** for **two sets** of correct responses.

Do not award any marks for only one set of correct responses.

Number – Fractions, decimals and percentages

Fractions of amounts (page 20)

1 a) 5 b) 15

Award **1 mark** for both answers correct.

2 £30 (Megan has £90, Ali has £60)

3 3

4 a) £600 b) £3,600

5 $\frac{3}{5}$

6 Award **2 marks** for the correct answer of 108 pages.

Award **1 mark** for evidence of the correct method with no more than one mathematical error, e.g.
72 pages ÷ 2 = 36 pages
36 pages + 72 pages = incorrect answer

Mixed numbers (page 21)

1 $\frac{20}{3}$

2 $1\frac{1}{4}, 1\frac{3}{4}, 2\frac{1}{4}, 2\frac{3}{4}, 3\frac{1}{4}$

3 $1\frac{1}{2}$ and $1\frac{4}{8}$

4 Award **2 marks** for all correct.

Award **1 mark** for any three correct.

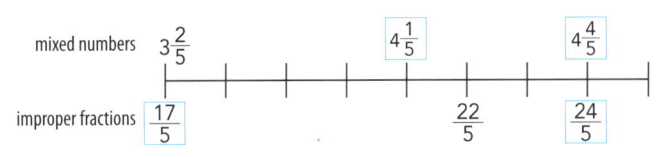

mixed numbers: $3\frac{2}{5}$, $4\frac{1}{5}$, $4\frac{4}{5}$

improper fractions: $\frac{17}{5}$, $\frac{22}{5}$, $\frac{24}{5}$

5 $\frac{30}{5}$ $\frac{18}{3}$ $\frac{54}{9}$

6 $2\frac{3}{4}$ or $\frac{11}{4}$

Equivalent fractions (page 22)

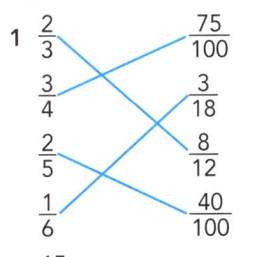

1 $\frac{2}{3}$ — $\frac{8}{12}$
$\frac{3}{4}$ — $\frac{75}{100}$
$\frac{2}{5}$ — $\frac{40}{100}$
$\frac{1}{6}$ — $\frac{3}{18}$

2 $\frac{15}{25}$

3 $\frac{1}{4}$ and $\frac{3}{4}$

4 $\frac{4}{100}$ m and $\frac{1}{4}$ m

5 $\frac{10}{20}$ $\frac{8}{20}$ $\frac{15}{20}$

6 $\frac{5}{8}$ (five triangles) shaded in any way

Adding and subtracting fractions (page 23)

1 $1\frac{2}{5}$

2 $\frac{4}{5}$

3 $\frac{9}{6}$ or $1\frac{3}{6}$ or $1\frac{1}{2}$

4 $\frac{3}{4}$ or any equivalent

5 Award **2 marks** for all correct.

Award **1 mark** for at least two correct.

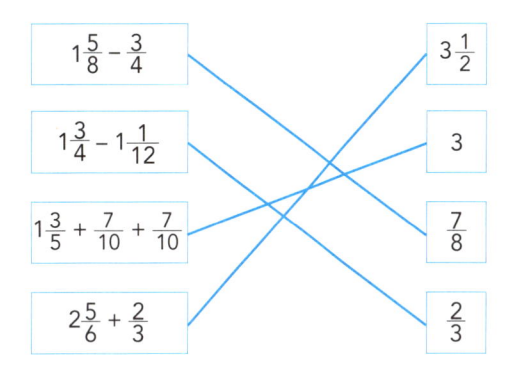

$1\frac{5}{8} - \frac{3}{4}$ — $\frac{7}{8}$

$1\frac{3}{4} - 1\frac{1}{12}$ — $\frac{2}{3}$

$1\frac{3}{5} + \frac{7}{10} + \frac{7}{10}$ — 3

$2\frac{5}{6} + \frac{2}{3}$ — $3\frac{1}{2}$

6 $1\frac{1}{8}$

Fractions and their decimal equivalents (page 24)

1 Award **2 marks** for all correct.

Award **1 mark** for four correct.

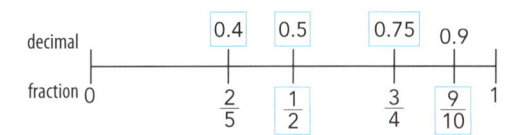

decimal: 0.4, 0.5, 0.75, 0.9

fraction: $\frac{2}{5}$, $\frac{1}{2}$, $\frac{3}{4}$, $\frac{9}{10}$

2 $\frac{36}{100}$

3 $\frac{25}{100}, \frac{5}{20}$ and $\frac{1}{4}$

4 0.375, 0.7, 0.43

Adding and subtracting decimals (page 25)

1 a) 44.15
b) 3.9

Award **1 mark** for both answers correct.

2 a) £35.24
b) £4.76

3 1.7 and 4.3; 2.46 and 3.54

4 2.11

5 Containers B and C (also award the mark for calculations that clearly show that B and C have been identified but have not been written in the answer boxes).

59

ANSWERS

Multiplying decimals (page 26)
1. 0.7
2. 6.4
3. 2
4. 4.7
5.
   0.6 × 3 = 1.8
   0.6 × 9 = 5.4
   0.6 × 7 = 4.2
   0.6 × 8 = 4.8

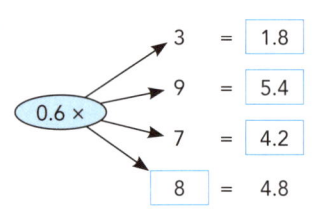

   Award **2 marks** for all four correct or **1 mark** for three correct.
6. 0.8 (accept $\frac{4}{5}$)
7. 12.5 m²

Percentages as fractions and decimals (page 27)
1. Award **2 marks** for all correct.
   Award **1 mark** for at least four correct.

Fraction	$\frac{3}{10}$	$\frac{65}{100}$ or $\frac{13}{20}$	$\frac{5}{100}$ or $\frac{1}{20}$	$\frac{4}{5}$
Decimal	0.3	0.65	0.05	0.8
Percentage	30%	65%	5%	80%

2. 1.4; 14%; 2.5
3. $\frac{7}{10}$, 70, 0.7
4. Award **2 marks** for all correct.
   Award **1 mark** for at least four correct.

   decimal: 0.1, 0.4, 0.5, 0.75, 0.95
   percentage: 0, 10%, 40%, 50%, 75%, 95%, 1

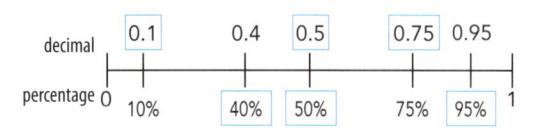

5. a) An explanation to show that either:
   - $\frac{2}{5}$ is equivalent to 40%, not 25%
   - 25% is equal to $\frac{1}{4}$ and this is not the same as (or is smaller than) $\frac{2}{5}$

   b) 0.4

Finding percentages (page 28)
1. a) 200   b) 400
   Award **1 mark** for both answers correct.
2. 24
3. £40
4. =
5. a) Award **2 marks** for 80 shown clearly on the bar chart.
   Award **1 mark** for a bar clearly identified but drawn incorrectly or missing.

   Do not award any marks for bar shown as 60 or less or 100 or more with no evidence of 80 identified in some way.

   b) 182

Ratio and proportion

Ratio and proportion (page 29)
1. 12
2. 400
3. Award **2 marks** for correct answer 72.
   Award **1 mark** for the scaling by 12 completed correctly, but with a mis-measurement and an answer (e.g. mis-measured as 5 cm and answer given as 60 km).
4. $\frac{17}{20}$
5. Red Team 100, Blue Team 60
   Award **2 marks** for both answers correct.
   Award **1 mark** for workings that clearly show 160 ÷ 8 = 20 as part of the calculation **and** one team's score is correct.

Algebra

Algebra (page 30)
1. 93 in top row, 5 in bottom row
2. 36; 35; 13
3. Area of a rectangle is length × width (breadth). May be written as base × height. Answer may be given as a formula, i.e. Area of rectangle = length × width or algebraically as A = l × w or A = lw (also accept a worked example instead of the formula).
4. a) Description of a = 180° − (100° + 35°) **or** 180° − 100° − 35°
   b) 45
5. a) 48
   b) q = 20

Sequences (page 31)
1. −8; −1; 6; 13; 20
2. The rule is subtract 20. Missing terms are 172, 132, 112
3. 188 and 68; the rule is subtract 60
4. −27
5. Award **1 mark** for NO, and an explanation that shows that since 7,125 will be in the sequence after 10 jumps of 700; 7,025 cannot be in the sequence.

   Do not award any marks for recognition that all numbers in the sequence will end in 25 as this could imply that it can be any number ending in 25.

Solving equations (page 32)
1. e.g. ☐3 + 15 = ☐6 × 3
2. 25 minutes and 35 minutes
3. 8 and 3, or 24 and 1
4. A = 750 g, B = 250 g

Measurement

Length (page 33)
1. 325
2. 18
3. 38 cm — 38 mm
   3.8 cm — 380 cm
   3.8 m — 0.38 m
   (crossed matching)

60

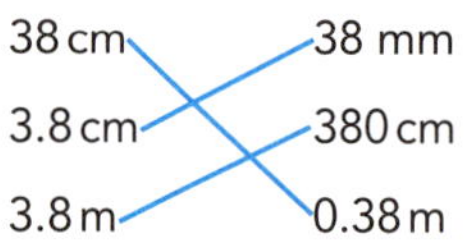

ANSWERS

4 5,600

5 0.316

6 Award **2 marks** for the correct answer of 4.32.

Award **1 mark** for evidence of the correct method with no more than one mathematical error, e.g.

7.2 cm × 60 = 432 cm

432 cm ÷ 100 = wrong answer

Mass (page 34)

1 4,310

2 Award **2 marks** for correct answer 1.625 kg or 1,625 g.

Award **1 mark** of correct reading of 875 g shown and evidence of a correct method,

e.g. 2,500 g – 875 g = error.

Do not award any marks for an answer of only 875 g with second step of problem omitted.

3 0.28

4 Award **2 marks** for correct answer 415.

Award **1 mark** for a correct/appropriate method

e.g. 5,500 g – 520 g = 4,980 g
     4,980 g ÷ 12 = error

or   5,500 g – 520 g = error
     error ÷ 12 = error

Do not award any marks for evidence of only one step in the problem.

Capacity (page 35)

1 4,650

2 Two amounts totalling 1,250

3 10

4 30

5 a) 10,000

   b) 5

Money (page 36)

1 a) £26.25

   b) £1.20

2 Award **2 marks** for correct answer hoody (£11.25) and cap (£6.99).

Award **1 mark** for evidence of total needing to be £18.24 but with error when adding two items.

Do not award any marks for only identifying £18.24 with no evidence of the second step.

3 Award **2 marks** for correct answer Kelsey £4.45 and Toby £3.20

Award **1 mark** for evidence of correct working but incorrect answers.

4 a) £18.75

   b) £66.25

5 Award **2 marks** for correct answer £52.50

Award **1 mark** for evidence of all steps of the problem.

Do not award any marks for an incomplete method.

Time (page 37)

1 Award **2 marks** for all correct.

Award **1 mark** for two correct.

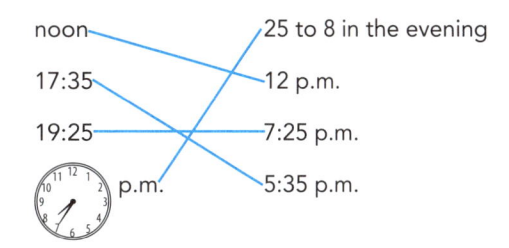

noon — 12 p.m.
17:35 — 5:35 p.m.
19:25 — 7:25 p.m.
(clock showing 7:35) — 25 to 8 in the evening

2 C (11:30 a.m.), D (12:00), B (21:35), A (10:40 p.m.)

3 8 o'clock or 8:00

4 11:45 a.m.

Time problems (page 38)

1 1 hour 45 minutes (accept 105 minutes)

2 13:50 or 1:50 p.m. (award the mark even if p.m. is not shown)

3 a) 10:30 or half past 10   b) £1.75

4 a) 1 hour 50 minutes (accept 110 minutes)

   b) 51 minutes

   c) 12:57 (accept 3 minutes to one)

Perimeter (page 39)

1 30

2 64.4

3 114

4 Award **2 marks** for correct answer 41.4 cm.

Award **1 mark** for 414 mm or for correct measurements of unlabelled sides identified but with error when calculating the perimeter.

Do not award any marks for correct measurements added to the diagram but with no signs of calculating the perimeter.

Estimating the area of irregular shapes (page 40)

1 Accept any shape with an area of $7\frac{1}{2}$ squares.

2 Shape B

3 Approximately $11\frac{1}{2}$ cm² (11.5 cm²) (accept an answer that is more than 11 cm² but less than $11\frac{3}{4}$ cm²)

4 Award **1 mark** for an explanation that shows in some way that a better estimate or approximation would be 17–17.5 m². This may be through annotations on the diagram that clearly show that the area is more than 15 m².

Do not award any marks for an explanation that Asha has not counted the squares carefully.

Area by formula (page 41)

1 121 cm²

2 A (36 m²)

3 Any two sets of dimensions using factors of 36 (1 cm and 36 cm, 2 cm and 18 cm, 3 cm and 12 cm, 4 cm and 9 cm) or more sophisticated responses such as 5 cm and 7.2 cm

4 Award **2 marks** for correct answer 3 m².

Award **1 mark** for evidence of a correct method e.g.
4 m × 3 m = 12 m²
3 m × 3 m = 9 m²
12 m² – 9 m² = error, or for giving the answer in cm²

61

ANSWERS

Geometry – Properties of shapes

Drawing lines and angles (page 42)

1. Rectangle drawn accurately with 90° angles and sides of 6.5 cm and 3.5 cm
2. a) 7.5 (accept 7.3 to 7.7)
   b) 35 and 55
   (give margin error of 2° on either side of exact measure)
3. a) Drawn to scale
   b) Approximately 110 to 115

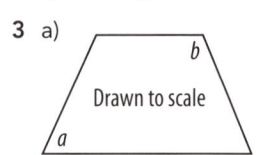

2-D shapes (page 43)

1.

Name of 2-D shape	Number of straight sides	Number of right angles
rectangle	4	4
right-angled triangle	3	1
regular hexagon	6	0
equilateral triangle	3	0

2. (heptagon, arrow, flag ticked)

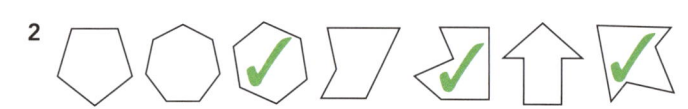

3. rectangle, rhombus, regular octagon

4.

Name of shape	Number of pairs of parallel lines	Number of pairs of perpendicular lines	Total number of sides
irregular pentagon	1	2	5

3-D shapes (page 44)

1. 5 faces, 9 edges, 6 vertices
2. Rectangular face: D; Both: B; Triangular face: C; Neither: A, E
3. square-based pyramid
4. C and E ticked
5. A and C ticked

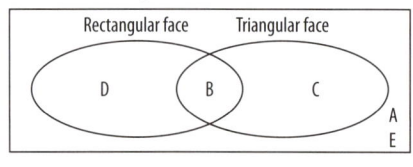

Angles and degrees (page 45)

1. 120
2. 50
3. $a = 72$ and $b = 135$
4. 75
5. Award **2 marks** for correct answer 105.
   Award **1 mark** for evidence of the correct method (i.e. 360 – 150 = 210, then divided by 2 = incorrect answer).
   Do not award any marks for not recognising that angles at a point sum to 360° or for omitting a step in the problem.

Angles in triangles (page 46)

1. 42
2. All angles labelled as 60° **or** sight of calculation 180° ÷ 3
3. $a = 90°$ $b = 55°$
4. 55° and 55°
5. $a = 90$ and $b = 45$

Geometry – Position and direction

Coordinates (page 47)

1. **a** (2,2), **b** (6,4), **c** (3,6)
2. (8,5)
3. pentagon (accept irregular pentagon)

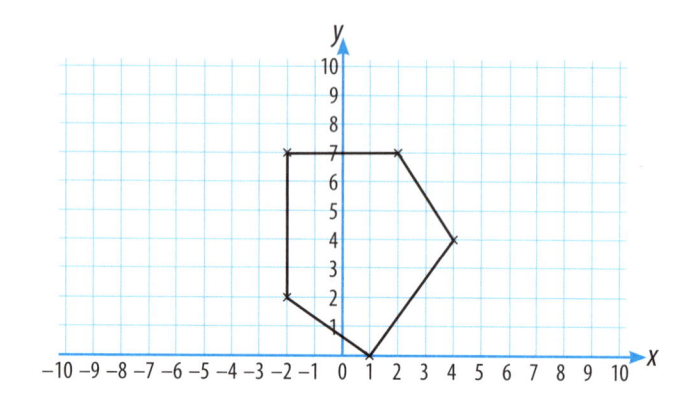

Award **2 marks** for a correctly named shape and correct shape drawn on grid as shown here.

Award **1 mark** for all coordinates correctly positioned but not joined or shape unnamed.

Do not award any marks if any coordinates are incorrectly positioned.

4. (5,7) and (8,7) **or** (5,1) and (8,1)

Translations (page 48)

1. (0,0)
2.

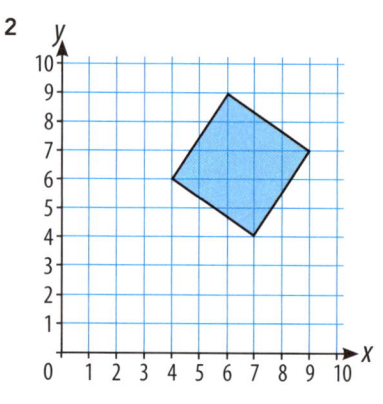

3. 2 squares right and 3 squares down

ANSWERS

Reflections (page 49)
1 C
2
3
4

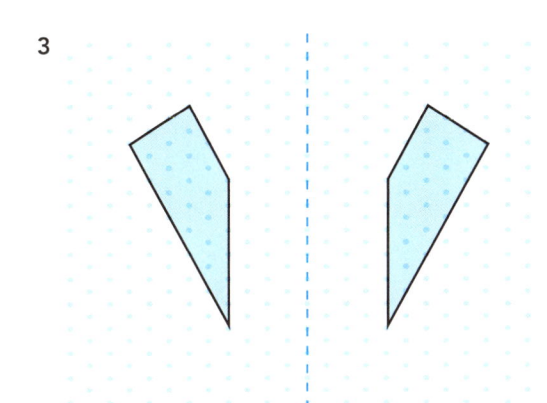

Reflective symmetry (page 50)
1
2
3

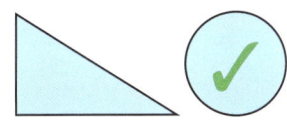

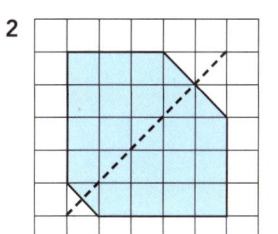

	Reflective symmetry	No reflective symmetry
Quadrilateral	E	C
Not quadrilateral	B	D, A

4

Statistics

Tables (page 51)
1

Favourite sport	Tally	Total
badminton	|||| |||| |||| |||| |||	23
hockey	|||| |||| ||	12
football	|||| |||| |||| |||	18
		53

2 a) £39.43
   b) £3.74

3 a) Wednesday b) Kate and Jayden

4 a) $1\frac{1}{2}$

   b) Award **2 marks** for correct answer $32\frac{1}{4}$.
      Award **1 mark** for evidence of a correct method.
      Do not award any marks for an incomplete method.

Pictograms (page 52)
1
Red
Blue
Green

= 3 points

ANSWERS

2 a) 10   b) 62   c) [diagram: two full circles and a quarter circle]

3 a) gardening and fashion
   b) sports and games

Bar charts (page 53)

1 a) 13   b) 103

2 a)

	Object A	Object B	Object C	Object D
Mass (g)	350	500	550	650

   b) Bar correctly drawn to show 450 g

3 a) 2
   b) Accept a bar that is clearly more than £10 but less than £12.50
   c) Award **2 marks** for the correct answer of 11.25.
      Award **1 mark** for evidence of the correct method with no more than one mathematical error, e.g.
      12.50 + 25 + 17.50 + 22.50 + 11.25 = 88.75
      £100 − £88.75 = wrong answer

Pie charts (page 54)

1 a) $\frac{3}{4}$ b) 90
   c) YES, supported by an explanation that shows clearly that $\frac{1}{4} + \frac{1}{4} = \frac{1}{2}$

2 a) $\frac{3}{8}$ b) £30   c) £30

Line graphs (page 55)

1 a)

Day	4	8	14	20
Height (mm)	10	25	35	55

   b) 67.5 (accept an answer that is greater than 66 but less than 69)

2 a) 5   b) 20   c) $2\frac{1}{2}$ hours

Averages (page 56)

1 9

2 Red = 6; Blue = 10; Yellow = 8
   Total = 24 Mean is 24 ÷ 3 = 8

3 $2\frac{1}{2}$

4 10

5 57.5

6 64 minutes
   Award **2 marks** for correct answer of 64 minutes.
   Award **1 mark** for correct method of total of Monday, Tuesday, Wednesday, Thursday and Friday divided by 5 but an incorrect answer.